Operation and Maintenance of Drones Used for Agricultural Applications

CONTENT VIBES
PUBLISHERS ▪ PRINTERS ▪ AGGREGATORS

About the Author

Tejas Ashok Bhosale completed his Bachelor of Technology (B. Tech.) from Mahatma Phule Krishi Vidyapeeth (MPKV), Rahuri, Maharashtra. He qualified for the Junior Research Fellowship (JRF) examination conducted by the Indian Council of Agricultural Research (ICAR). Subsequently, he earned his Master of Technology (M. Tech.) from G.B. Pant University of Agriculture and Technology (GBPUAT), Pantnagar. He has also qualified for the ICAR Agricultural Scientists Recruitment Board National Eligibility Test (ICAR-ASRB-NET). With over 12 years of teaching experience in State Agricultural Universities (SAUs), Tejas is currently serving as an Assistant Professor at Veer Chandra Singh Garhwali Uttarakhand University of Horticulture and Forestry (VCSGUUHF), Bharsar, Pauri Garhwal, Uttarakhand. He is a DGCA-certified drone pilot, specializing in the application of drones within the agricultural sector. Tejas has published 7 research papers, 5 popular articles, 1 book chapter, and 1 book. He has handled one project as Principal Investigator (PI) and another as Co-Principal Investigator (Co-PI). Additionally, he possesses administrative experience in various capacities within SAUs. His areas of specialization include agricultural mechanization, precision agriculture, the use of drones in agriculture, and the design and development of intercultural and harvesting machinery. Tejas's expertise and contributions significantly enhance agricultural technology and education.

Operation and Maintenance of Drones Used for Agricultural Applications

Tejas Ashok Bhosale
Assistant Professor
Agricultural Engineering
Department of Natural Resource Management
College of Horticulture
Veer Chandra Singh Garhwali Uttarakhand University of Horticulture and Forestry
Bharsar, Pauri Garhwal, Uttarakhand

CV CONTENT VIBES
PUBLISHERS • PRINTERS • AGGREGATORS

CONTENT VIBES
PUBLISHERS • PRINTERS • AGGREGATORS

1168, Sector 13, Urban Estate
Karnal-132 001, Haryana
Tel: 91-84470 75807, 18440 41168
Email: contentvibesppa@gmail.com
www.contentvibes.in

Print ISBN: 978-93-67550-55-7

ebooks ISBN: 978-93-67558-23-2

Preface

The evolution of drone technology has brought about revolutionary changes across various industries, and agriculture is no exception. "Operation and Maintenance of Drones Used for Agricultural Applications" delves deep into the role that unmanned aerial vehicles (UAVs) play in modern farming practices. This book is designed for farmers, agronomists, and drone enthusiasts who seek to explore the exciting intersection of technology and agriculture. It offers both a comprehensive introduction to drone technology and an in-depth understanding of how these tools can be utilized to enhance productivity and sustainability in agriculture.

The book is organized into 13 chapters, each focusing on a specific aspect of drone technology and its application in agricultural settings. Starting with an introduction to the technology behind drones, it progresses into the various components that make up a drone, from motors and propellers to flight control systems and sensors. The heart of the book explores how drones are equipped with advanced sensors such as multispectral, thermal, and LiDAR cameras, which are crucial in tasks like crop monitoring, pest detection, and soil moisture assessment.

One of the primary goals of this book is to provide readers with a holistic view of drone operation, from flight planning and data collection to routine maintenance and safety protocols. Special attention is given to the practical applications of drones in crop monitoring, early pest detection, irrigation management, and overall farm management. The book also highlights the regulatory landscape surrounding drone operations, ensuring that readers are well-versed in the legalities of drone use in agriculture.

Ultimately, "Operation and Maintenance of Drones Used for Agricultural Applications" aims to empower its readers with the knowledge needed to adopt and effectively use drones, helping them navigate the future of agriculture with efficiency, precision, and innovation.

—Author

Contents

1

Drone Technology

OVERVIEW OF UNMANNED AERIAL VEHICLE (UAV) TECHNOLOGY

Agricultural drone technology employs drones equipped with cameras and sensors to assist farmers in monitoring crops, applying pesticides, and analyzing soil conditions. This technology assists farmers in making informed decisions, enhancing yields, and minimizing chemical application.

A drone is fundamentally a flying robot that can be operated remotely or autonomously through software-driven flight plans within its embedded systems, which integrate with onboard sensors and a Global Positioning System (GPS).

Drones are remotely operated or autonomous aircraft engineered to function without a pilot on board. They are also referred to as Unmanned Aerial Vehicles (UAVs). They has the capability to function alone or be managed by human operators.

- They have gained significant popularity in recent years owing to their extensive range of applications and uses. Drones are available in a range of sizes and configurations, from compact portable models to larger variants. They possess an array of equipment that enables optimal functionality.
- The fundamental components of a drone consist of motors, propellers, flight controllers, batteries, and remote control systems. The motor and propeller are essential for generating push to elevate the drone from the ground, whilst the flight controller functions as the drone's brain, responsible for data processing and movement regulation.
- The battery energizes the drone's electronic components, while the remote control systems enable the operator to maneuver the drone. The drone functions by an integration of electronic components and software algorithms.

Figure: 1.1 Assessing Agri Drone Viability in India

Drones are utilized across several applications and sectors, including agricultural, military operations, photography and videography, delivery services, and search and rescue missions. Drones are utilized in agriculture for crop monitoring and in the military for surveillance purposes.

Drones capture aerial images in photography and cinematography, and are utilized for product delivery in distant regions. Search and rescue operations are employed to locate missing persons and facilitate catastrophe management.

WORKING OF A DRONE

- Drones operate by an integration of physical components and software algorithms. The drone operation commences with the flight controller, which regulates the drone's movement, stability, and direction in the air. The flight controller supplies electricity to the drone's motors, propellers, and additional components.
- The majority of drones utilize rotors and propellers to produce power and elevate themselves from the ground. Multi-rotor drones, including quadcopters, hexacopters, and octocopters, include many rotors configured in symmetrical patterns to ensure adequate lift and control. Upon receiving electricity, the drone motors initiate rotation, elevating the drone into the air.

- Upon ascending, the flight controller identifies the flying level while the GPS supplies data regarding the drone's location, enabling the pilot to maneuver the drone's trajectory and motions.
- Subsequently, supplementary algorithms are employed to guide drones around barriers and prevent collisions with real-time entities. Communication technologies enable drones to receive directives from a human operator via a remote controller or base station. Certain drones interact with other drones or gadgets inside a networked setting to facilitate collaborative missions or swarm behavior.

Drones are equipped with safety features including return-to-home functions, automated landing capabilities, and fail-safe systems to guarantee safe operation and mitigate the risk of accidents or malfunctions.

Usage of Sensors Agricultural Drones

A variety of sensors are installed in agricultural drones, few of which are listed below-

- **Multispectral Cameras:** These sensors, which collect data across multiple wavelengths, enable farmers to assess plant health, recognize early disease indicators, and identify nutritional deficiencies.
- **RGB Cameras:** These sensors collect high-resolution images of crops, providing farmers with detailed information regarding crop health, growth rates, and issues such as pest infestations.
- **Multispectral Cameras:** These sensors enable farmers to assess plant health, recognize early signs of disease, and pinpoint nutritional deficiencies. They collect data over multiple wavelengths.
- **Thermal Cameras:** These sensors identify heat signatures, providing farmers with critical information regarding pest infestations, plant stress, and irrigation needs.
- LiDAR sensors enable farmers to analyze topographical data, identify crop patterns, and strategize for more efficient planting by utilizing lasers to generate three-dimensional maps of the terrain.
- **GPS:** This sensor enables drones to automatically map fields, monitor agricultural growth, and identify potential issues.

Drones Utilized in Agriculture

Drones are beneficial for soil and field analysis throughout the initial stages of the agricultural cycle. They produce precise three-dimensional maps for preliminary soil analysis, which might inform seed planting designs. Drone-assisted soil analysis provides data for the management of nitrogen levels and irrigation post-planting.

- **Planting:** Emerging drone-planting technologies have a 75% adoption rate and reduce planting costs by 85%. These systems provide plants with essential elements for survival by deploying pods laden with nutrients into the soil.

Drones utilized for crop spraying are equipped with distance-measuring systems, including ultrasonic echoes and lasers employed in the light-detection and ranging (LiDAR) approach, enabling them to adjust their altitude in accordance with variations in topography and geography, so averting collisions. Drones may modify their altitude and apply liquid in real time for consistent coverage, simultaneously surveying the terrain and dispensing the correct volume of liquid. Drones can minimize chemical usage and its adverse environmental impacts by administering insecticides and fertilizers with greater precision than traditional techniques.

Drones enable farmers to oversee their crops from an aerial perspective, gathering data on crop vitality, pest and disease prevalence, and irrigation requirements. This information can facilitate educated decisions regarding crop management, including the timing of fertilizer or insecticide application.

Drones outfitted with thermal, multispectral, or hyperspectral sensors can identify portions of a field that require enhancement or are arid. Drones facilitate the calculation of the vegetation index, reflecting the crop's relative density and health, and reveal the crop's heat signature, indicating the amount of heat or energy emitted during growth.

- **Mapping and Imaging Applications:** Throughout the manufacturing process, the mapping and imaging functionalities of drone platforms, equipped with various sensors, can enhance production planning and efficiency. Drone technology can be employed to evaluate soil conditions prior to the commencement of the growth cycle in order to estimate prospective yields. The most crucial application for evaluating soil quality is the accurate 3D mapping of the topography, featuring precise soil color coverage. This aids in accurately assessing soil quality, moisture levels, and water movement within the soil. During the vegetative phase, planners can utilize cyclic flights to oversee crops and the agricultural process, allowing them to organize activities and swiftly respond to emerging difficulties.
- **Health Assessment:** Assessing crop vitality and detecting fungal or bacterial illnesses in trees is essential. Drone-mounted apparatus can identify plants that reflect differing amounts of green light and near-infrared light by scanning crops with both visible and NIR light. This data can be utilized to produce multispectral images that monitor alterations in vegetation and indicate their vitality. Furthermore, farmers can more precisely administer and oversee medications immediately upon the identification of a disease.

  Drones can accurately detect and target weeds utilizing high-resolution sensors and advanced AI technologies, thereby obviating the necessity for indiscriminate pesticide sprays. This targeted technique minimizes herbicide usage, conserves financial resources for farmers, and enhances overall yields. Drones equipped with sprayers can immediately apply herbicides to the targeted weeds. This reduces drift and diminishes the likelihood of herbicide harm to adjacent crops.

Advantages of Agri-Drones

Security

The drones are controlled by certified drone operators. Consequently, the likelihood of their misuse is nonexistent.

Exceptional Efficiency

Drones operate without delays and can function at twice the speed of human labor.

Water Conservation

Compared to conventional spraying techniques, agricultural drones employ ultra-low volume (ULV) spraying technology, resulting in greater water conservation.

Economical and simple to Sustain

Agricultural drones are robust, economical, and necessitate minimal upkeep. Notable features encompass a detachable container, an economical frame, and accurate pesticide application.

Constraints of Agri Drones

Connectivity Problem

Frequently, online access is lacking in remote regions. on these conditions, a farmer must invest on internet connectivity, which may become a continual expense.

Contingent Upon Weather Conditions

Drones have no operating delays and can function at twice the speed of human labor.

Contingent Upon Weather Conditions

Drones are significantly reliant on favorable weather conditions. It is inadvisable to operate drones during inclement weather, such as rain or strong winds.

Expertise and Proficiency

Adopting new technology is a positive development; yet, daily utilization necessitates the appropriate skill set and sufficient understanding. A typical farmer may find it challenging to comprehend drone operations. He must either acquire information or stay reliant on an experienced someone.

Emeaging Trends and Innovations in Drone Technology

- **Advanced AI and Machine Learning:** Drones are being incorporated into smart agriculture through the combination of automation, machine learning, and artificial intelligence. Drones integrated with AI software will provide farmers with extensive insights as data analytics advance, including yield pattern forecasting and ideal planting timing recommendations. The intelligent agricultural ecosystem will promote a proactive farming strategy through data-driven decision-making, resource optimization, and a focus on sustainability.

- **Drone Swarming:** Innovative applications, such as coordinated activities performed by swarming drones, are anticipated for the future. Drone swarms can swiftly traverse extensive areas, performing tasks such as pollination, collaborative planting, and agricultural surveillance. This collaborative strategy expedites the implementation of innovative agricultural methods and enhances efficiency.
- **Integration with Other Technologies:** Agricultural drones can be integrated with advanced technologies such as blockchain, Internet of Things (IoT) devices, and satellite imagery. This integration will enable farmers to comprehensively comprehend their operations and establish a resilient ecosystem.
- **Enhanced Integration of Multispectral Sensors:** Agricultural drones are progressively employing multispectral sensors. These sensors capture data across many spectrums, including infrared, facilitating a more comprehensive and accurate analysis of crops. Farmers can assess plant health, identify illnesses, and administer fertilizers with optimal efficiency through multispectral imaging.
- **Customization for Specialized Agriculture:** Farmers' requirements differ according to the types of crops cultivated and the specific challenges encountered. Manufacturers are offering customizable drones to address diverse agricultural requirements. Specialized drones are designed to address the distinct requirements of various agricultural settings, including vineyards, orchards, and expansive crop fields.
- **Augment Data-Analytics Services:** This encompasses not merely the collection of data, but also the extraction of valuable insights. An increasing number of agricultural drone firms are offering data analytics services. These programs aid farmers in interpreting drone data and converting it into actionable recommendations for optimizing crop management.
- **Drone-as-a-Service (DaaS) Models:** The increasing prevalence of drone-as-a-service (DaaS) models enables farmers to utilize drone technology without incurring the upfront costs associated with purchasing and maintaining their own drones. DaaS providers facilitate broader access to drone technology for farmers by offering specialized drone services for certain tasks.

SIGNIFICANCE OF DRONES IN AGRICULTURE

This text discusses existing applications of drone technologies in agriculture, explores emerging agricultural drone innovations, and addresses the stages and issues associated with the mainstream adoption of drones in farming.

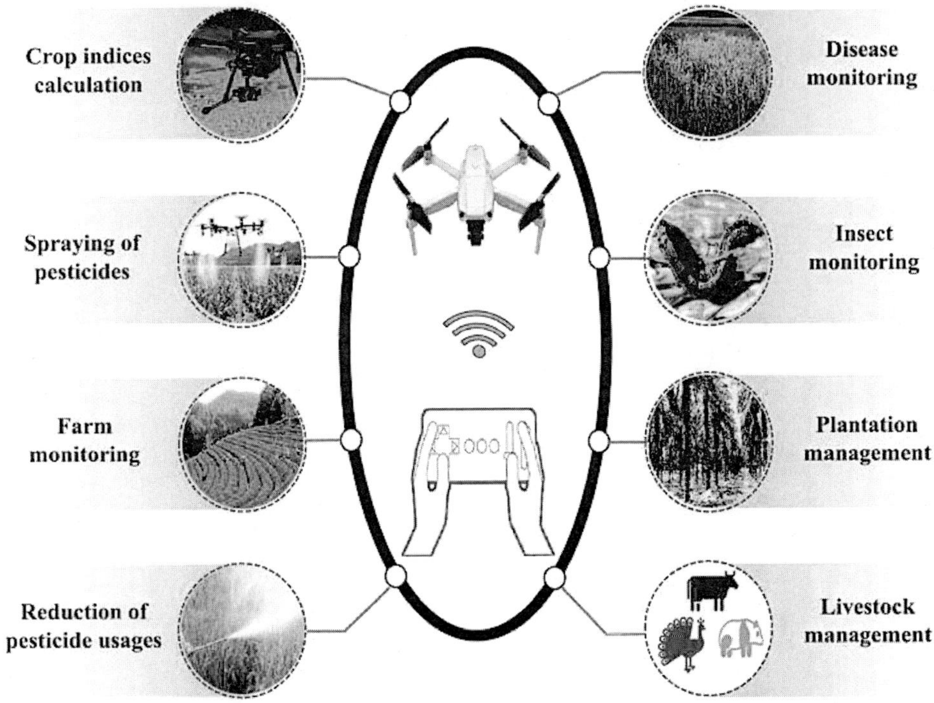

Figure: 1.2 Utilization of Drone Technology in Agriculture

Scouting/Monitoring Plant Health

A successful application of drone imaging is the monitoring of plant health. Drones outfitted with specialized imaging technology known as Normalized Difference Vegetation Index (NDVI) utilize intricate color data to assess plant vitality. This enables farmers to oversee crop development, facilitating prompt intervention to mitigate issues and preserve the plants. This graphic succinctly demonstrates the functionality of NDVI.

Drones equipped with conventional cameras are utilized to assess crop health. Numerous farmers currently utilize satellite images to assess crop growth, density, and coloration; however, obtaining satellite data is expensive and often less effective than proximal drone imaging. Drones operate near fields, rendering cloud cover and inadequate lighting conditions less significant compared to satellite photography. Satellite imaging can achieve accuracy to the meter, whereas drone imaging can provide precise picture placement to the millimeter. This indicates that post-planting, locations with gaps in the stand can be identified and replanted as necessary, while disease or pest issues can be promptly diagnosed and addressed.

A successful application of drone imaging is the monitoring of plant health. Drones outfitted with specialized imaging technology known as Normalized Difference Vegetation Index (NDVI) utilize intricate color data to assess plant vitality. This enables farmers to oversee crop development, facilitating prompt intervention to mitigate issues and preserve the plants. This graphic succinctly demonstrates the functionality of NDVI.

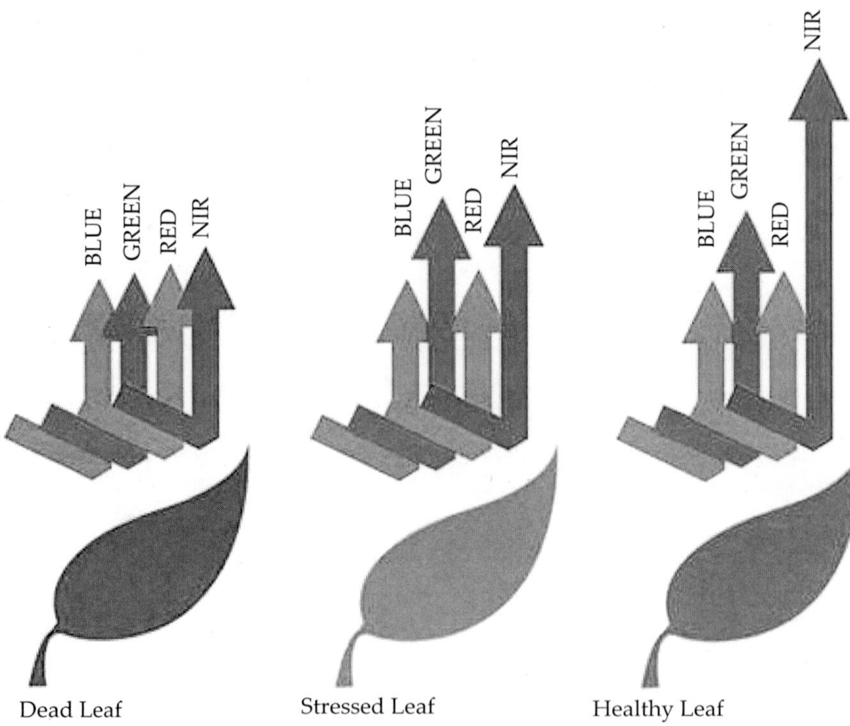

Figure: 1.3 Assessment and Surveillance of Plant Health

Drones equipped with conventional cameras are utilized to assess crop health. Numerous farmers currently utilize satellite images to assess crop growth, density, and coloration; however, obtaining satellite data is expensive and often less effective than proximal drone imaging. Drones operate near fields, rendering cloud cover and inadequate lighting conditions less significant compared to satellite photography. Satellite imaging can achieve accuracy to the meter, whereas drone imaging can provide precise picture placement to the millimeter. This indicates that post-planting, locations with gaps in the stand can be identified and replanted as necessary, while disease or pest issues can be promptly diagnosed and addressed.

Monitoring Field Conditions

Drone-based field monitoring is employed to assess soil health and field conditions. Drones can deliver precise field mapping, including elevation data, enabling producers to identify any anomalies in the area. Knowledge of field elevation is essential for identifying drainage patterns and wet/dry areas, facilitating more efficient irrigation methods. Certain agricultural drone retailers and service providers additionally give nitrogen level monitoring in soil utilizing advanced sensors. This facilitates the accurate application of nutrients, eradicating suboptimal growth areas and enhancing soil health for the long term.

Soil and Field Examination

Agricultural drones can be utilized for soil and field analysis to enhance field planning efficiency. They can be utilized to affix sensors for assessing moisture levels, topographical conditions, soil characteristics, erosion, nutrient content, and soil fertility.

Planting & Seeding

Agri-drones are equipped with reservoirs that can be filled with fertilizers and pesticides, enabling rapid application on crops, much outpacing previous methods. Consequently, drone technology has the potential to initiate a new epoch in precision agriculture.

Figure: 1.4 The Utilization of Drones for Spray

Application of Spray

The utilization of drones for spray applications is prevalent in Southeast Asia, with South Korea employing drones for about 30% of its agricultural spraying activities. Drone sprayers may access extremely inaccessible locations, such as steep tea plantations at elevated altitudes. Drone sprayers alleviate the necessity for personnel to traverse fields using backpack sprayers, which pose health risks. Drones equipped with sprayers give precise spray treatments that may be directed to specified regions, optimizing efficiency and reducing chemical expenditures. Presently, regulations governing drone sprayers differ significantly across nations. In Canada, they are presently illegal due to the necessity for further testing to comprehend the effects of spray drift. Certain regulatory suggestions advocate that only qualified professionals operate spray drones, similar to Yamaha, which does not sell its manufactured spray drones but instead leases drone services with licensed operators.

Figure: 1.5 The Utilization of Drones for Security in Agricultural Farm

Security

The drone security sector is rapidly expanding beyond agriculture and is also highly beneficial for farm management. Employing drones to surveil remote sections of a farm eliminates the need for physical access, so conserving time and facilitating more frequent observation of inaccessible parts. Drone cameras offer a comprehensive perspective of agricultural activities throughout the day, facilitating the monitoring of operational efficiency and the identification of utilized equipment. Security drones may be utilized to surveil fencing and perimeters of high-value crops such as cannabis, rather than hiring additional security staff. Drone cameras are utilized innovatively to safeguard livestock by identifying missing or injured animals in remote grazing regions. Surveillance of remote regions, previously requiring hours of trekking, can now be accomplished in within minutes.

Figure: 1.6 The utilization of drones for Pollination

Drone Pollination

Several recent applications of drones in agriculture remain in the testing and research phase. A prominently advertised, if much dramatized, application is drone technology for pollination. Researchers in the Netherlands and Japan are creating diminutive drones that can pollinate plants without causing harm. The subsequent phase involves the development of autonomous pollinating drones capable of functioning and monitoring crop health independently, without continuous guidance from operators.

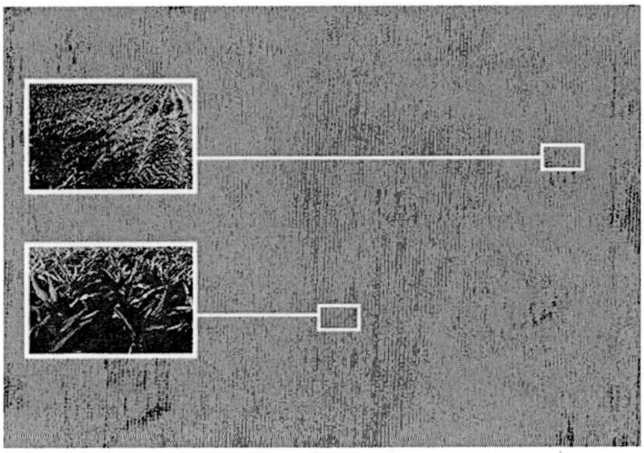

Figure: 1.7 Improving Artificial Intelligence (AI)

Drone AI

Another drone technology in development incorporates machine learning. Enhancing Artificial Intelligence (AI) in drones is crucial for increasing their utility for smallholder farmers in poor countries. Contemporary drone technologies are more efficient in surveilling established crops such as maize, which are cultivated in extensive monocultural field arrangements. Current drone monitoring programs struggle to identify regions with heightened crop diversity, lesser-known food, and grains that exhibit similar appearances during various growth stages, hence diminishing their efficacy in assessing crop growth and health. Additional efforts are required to enable AI systems to identify less prevalent crops and a broader array of planting patterns.

Figure: 1.8 The utilization of Drones for Irrigation

Drone Irrigation

Recent study from Australia is generating promising prospects for the application of drones in agriculture. As climate change progressively exacerbates drought conditions, the development of more effective irrigation technologies is essential. Drones utilize microwave sensors to provide precise soil health data, including moisture levels, unobstructed by vegetation. This indicates that water may be allocated on a field in the most efficient manner to conserve resources.

Livestock Management

Drones can monitor and manage extensive livestock due to their high-resolution infrared cameras, which can identify sick animals and facilitate prompt interventions. The influence of drones on precision dairy production is poised to become standard practice.

VARIANTS OF AGRICULTURAL DRONES (FIXED-WING, ROTARY-WING, MULTI-ROTOR)

These drones are employed for agricultural tasks including crop monitoring, insect identification, crop spraying, and precision irrigation. This assists farmers in achieving superior yields and executing high-quality agricultural practices.

Drones are referred to as unmanned aerial vehicles (UAVs). Drones are available in diverse types, each designed for particular applications and sectors based on their distinct capabilities and characteristics.

Various types of drones are frequently utilized in agriculture. Each type possesses distinct capabilities and traits that render them appropriate for various activities.

The following are prevalent categories of drones utilized in agriculture:

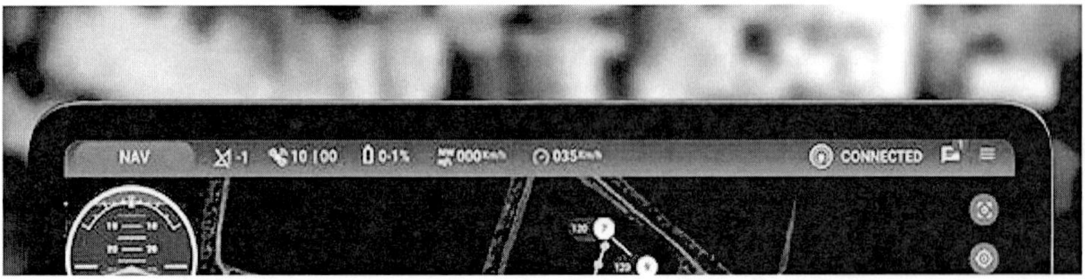

Figure: 1.9 Categories of Drones

Fixed-Wing Drones

These drones resemble conventional airplanes with wings. They exhibit efficiency in flight and traverse extensive distances. Fixed-wing drones are typically employed for aerial mapping, surveillance, and agricultural purposes.

Fixed-wing drones are a kind of unmanned aerial vehicles (UAVs) that resemble small airplanes. In contrast to multi-rotor drones, they include a fixed-wing configuration that enables prolonged aerial endurance and extensive coverage. They are generally launched manually and operated via remote control or a pre-set flight plan. Fixed-wing drones are frequently employed in agriculture for field mapping, crop monitoring, and detecting possible difficulties such as pests or irrigation deficiencies. They are additionally employed for surveying, cartography, and aerial photography. Fixed-wing drones are favored for precision agriculture applications due to their extended flight duration and capacity to traverse expansive areas. Nonetheless, they provide additional challenges in flight compared to multi-rotor drones, owing to their increased complexity and requirement for a more expansive area for takeoff and landing.

Figure: 1.10 A Diagram of Fixed-Wing Drone

Usage of Fixed-Wing Drones in Agriculture

1. Fixed-wing drones are frequently employed in agriculture for mapping, surveying, and crop monitoring activities. The applications of fixed-wing drones in agriculture encompass:
2. Acquiring high-resolution aerial photos of agricultural area for crop health assessment
3. Mapping domains to discern discrepancies in soil vitality and nutrient concentrations
4. Observing the growth and development of crops over time
5. Detection of pest and disease outbreaks in crops
6. Executing topographic assessments of agricultural land to facilitate irrigation planning

Merits and Demerits of Using Fixed-Wing Drones

Fixed-wing drones, like any technology, possess advantages and disadvantages in agricultural applications. Below are several Merits and Demerits of employing fixed-wing drones:

Merits

1. Can encompass extensive areas rapidly and effectively
2. Capable of sustained flight owing to their aerodynamic design
3. Capable of being outfitted with high-resolution cameras and sensors for comprehensive data acquisition.
4. Capable of functioning in inclement weather conditions
5. Exhibit a longer lifespan than multi-rotor drones owing to its more straightforward construction.

Demerits

1. Necessitate an expansive unobstructed space for takeoff and landing
2. May be more challenging to control than multi-rotor drones because to their dependence on a runway.
3. Incapable of hovering or executing rapid modifications to flight trajectory as multi-rotor drones can.
4. May necessitate increased maintenance compared to multi-rotor drones owing to their intricate construction.

Single-Rotor Drones

These drones feature a single rotor and a tail rotor to enhance stability and control. Single-rotor drones are less favored than multi-rotor drones owing to their elevated cost and intricate design.

Single-rotor drones, or helicopter drones, are unmanned aerial vehicles (UAVs) utilized in agriculture to obtain high-resolution photographs and data for crop mapping and analysis. They possess a singular rotor blade and are capable of vertical takeoff and landing, enabling them to hover and maneuver with greater precision than fixed-wing drones. Single-rotor drones are often larger and costlier than multi-rotor drones; however, they provide extended flight durations and the capacity to transport greater payloads, rendering them suitable for advanced applications such as precision agriculture. They are moreover outfitted with sophisticated sensors and cameras to gather data on crop vitality, soil conditions, and various environmental aspects. Single-rotor drones serve as an essential instrument for farmers and agricultural researchers aiming to enhance crop yields and efficiency while minimizing expenses and environmental effect.

Usage of Single-Rotor Drones in Agriculture

Single-rotor drones are generally employed for more specialized and exacting operations compared to fixed-wing drones. They are frequently utilized in agriculture for:

1. Agricultural surveillance and assessment
2. Irrigation administration
3. Soil examination and cartography
4. Quantification of flora and evaluation of health status
5. Application of pesticides and fertilizers in limited regions

Merits and Demerits of Using Single-Rotor Drones

Merits

1. More maneuverable and nimble than fixed-wing unmanned aerial vehicles.
2. More appropriate for small or unevenly shaped areas
3. Capable of maintaining a stationary position for prolonged durations, facilitating enhanced data acquisition.

Demerits

1. Restricted flight duration as comparison to fixed-wing drones
2. More intricate to run and sustain
3. Costlier than fixed-wing drones

Multi-Rotor Drones

Multi-rotor drones encompass quadcopters, hexacopters, and octocopters. These drones are configured in symmetrical patterns comprising numerous units. They exhibit exceptional efficiency and possess the ability for vertical takeoff and landing. These drones are utilized for aerial photography, videography, and inspection purposes. Quadcopter motors are engineered to provide ideal thrust and stability, facilitating smooth flight and precise maneuverability in these drones.

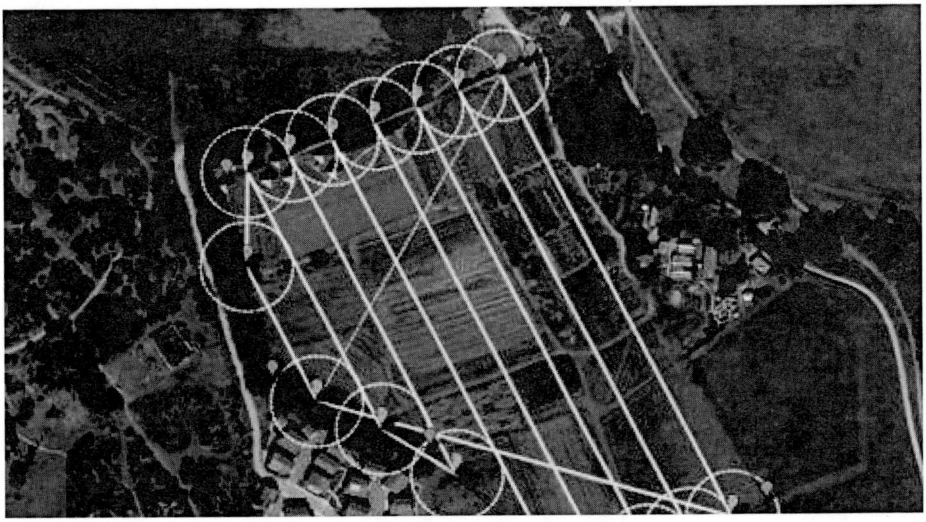

Figure: 1.11 Multi-Rotor Drones

Multi-rotor drones are characterized by several rotors affixed to their structure and are widely utilized in agriculture. These drones have exceptional maneuverability and can navigate in any direction, rendering them suitable for close-range tasks such as crop monitoring and analysis. Multi-rotor drones are frequently employed for mapping and surveying extensive agricultural regions, delivering high-resolution airborne data that facilitates the assessment of crop health, identification of irrigation problems, and monitoring of pest and disease outbreaks. They may also be outfitted with sophisticated sensors, like thermal cameras, to identify temperature fluctuations in crops, thereby assisting farmers in recognizing plant stress and modifying irrigation and fertilization practices accordingly. Multi-rotor drones possess the capability to hover, facilitating more accurate data collecting and monitoring. Nonetheless, they possess reduced flying durations and shorter ranges relative to fixed-wing drones, which may constrain their effectiveness in extensive agricultural enterprises.

Usage of Multi-Rotor Drones in Agriculture

Multi-rotor drones are adaptable and frequently employed in agriculture for a range of purposes, including:

1. Monitoring and mapping of crops
2. Quantification of flora and evaluation of health status
3. Soil examination
4. Management of irrigation
5. Application of pesticides and fertilizers in limited regions
6. Monitoring of livestock

Merits and Demerits of Using Multi-Rotor Drones

Merits

1. Simple to run and sustain
2. Capable of hovering still and maneuvering in confined areas
3. Less expensive than fixed-wing drones
4. Capable of transporting payloads including cameras or sensors

Demerits

1. Reduced flying durations in comparison to fixed-wing drones
2. Restricted range and velocity
3. Inappropriate for extensive areas or long-distance surveying

Hybrid Drones in Agriculture

Hybrid VTOL drones integrate fixed-wing and multi-rotor drone technologies. They possess vertical take-off and landing capabilities in conjunction with efficient forward flight. They are appropriate for aerial mapping, surveillance, and cargo transportation.

Hybrid drones in agriculture amalgamate the benefits of both multirotor and fixed-wing drones. These drones possess vertical take-off and landing (VTOL) characteristics akin to multirotor drones, together with the expanded range and endurance characteristic of fixed-wing drones. The hybrid design enables these drones to take off and land vertically, facilitating their deployment in restricted locations or rough terrain, while also permitting horizontal flight for prolonged durations, thereby covering extensive agricultural regions.

Hybrid drones are outfitted with several sensors and cameras that enable the collection of extensive data regarding crops and agricultural land. They can acquire high-resolution photos and produce precise maps of agricultural land, offering farmers critical information about the health and status of their crops. They can also be utilized to assess soil moisture levels, identify crop stress, and pinpoint parts of the farm that may want supplementary irrigation or fertilization.

Figure: 1.12 Hybrid Drones in Agriculture

The enhanced range and durability of hybrid drones render them suitable for extensive agricultural operations. They can encompass extensive agricultural regions in a single flight, diminishing the necessity for several flights and enhancing efficiency. Moreover, their prolonged flight duration enables lengthy aerial surveillance, facilitating the monitoring of crops and the identification of any alterations in their condition.

A further benefit of hybrid drones in agriculture is their capacity to function in a broader spectrum of meteorological circumstances. In contrast to fixed-wing drones that necessitate a runway or launch site, hybrid drones are capable of vertical takeoff and landing, rendering them less vulnerable to weather or uneven terrain.

Defining Hybrid Drones

Hybrid drones are unmanned aerial vehicles that integrate the characteristics of fixed-wing and multi-rotor drones. They possess the capability to ascend and descend like to a helicopter while exhibiting flight characteristics similar to those of an airplane. Hybrid drones utilize a combination of electric motors and gasoline engines to enhance range and flying duration.

Usage of Hybrid Drones in Agriculture

Hybrid drones are employed in agriculture for multiple functions, such as crop mapping, crop surveying, and pesticide application. The capability for vertical takeoff and landing, along with the range and velocity of a fixed-wing drone, renders hybrid drones optimal for extensive agricultural enterprises.

Applications of hybrid drones in agriculture include:

1. Detailed mapping at high resolution
2. Evaluation and surveillance of crop health
3. Management of irrigation
4. Precision agriculture
5. Enumeration of flora and assessment of yield
6. Application of pesticides

Merits and Demerits of Using Hybrid Drones in Agriculture

Merits

1. Hybrid drones provide superior range and flying duration compared to fixed-wing or multi-rotor drones.
2. They are capable of vertical takeoff and landing, hence obviating the necessity for a runway.
3. Hybrid drones provide the capability to transport greater payloads, rendering them suitable for agricultural spraying applications.
4. Hybrid drones deliver high-resolution pictures for precise mapping and agricultural analysis.

Demerits

1. Hybrid drones are costlier than fixed-wing or multi-rotor drones.
2. They are more intricate and necessitate greater upkeep than other drone varieties.
3. Hybrid drones necessitate particular training for safe and effective operation.
4. The utilization of gasoline engines in hybrid drones might generate noise and air pollution.

Hybrid drones represent the most sophisticated category of drones utilized in contemporary agriculture. They have more flexibility, range, and flight duration compared to other drone varieties, rendering them optimal for extensive agricultural enterprises. Despite their higher cost and complexity compared to other drone kinds, the advantages they provide render them a valuable investment for several farmers and agricultural enterprises.

Role of GIS in Agricultural Drone Technology

- **Remote Sensing:** GIS-enabled drones utilize remote sensing technologies to acquire data via diverse sensors, including RGB and multispectral cameras, thermal imaging, and LiDAR. These sensors gather information on crop health, soil moisture, and topography, facilitating the development of detailed maps for informed decision-making.
- **Spatial Analysis:** GIS enables spatial analysis by processing and analyzing data gathered by drones, which includes mapping and monitoring crop conditions, identifying regions of stress or disease, and evaluating the general health of the agricultural environment. Spatial analysis enables farmers to implement tailored interventions, optimizing resource allocation and enhancing production.

- **Precision Agriculture:** The amalgamation of GIS and drone technology facilitates precision agriculture practices, allowing farmers to accurately map differences in soil composition, moisture levels, and crop health, hence enabling the implementation of site-specific management strategies. This focused strategy diminishes resource waste, lessens ecological impact, and improves overall agricultural efficacy.
- **Data Visualization:** GIS technology facilitates the representation of drone-generated data through maps, charts, and 3D models, thereby enhancing the interpretation of intricate agricultural information and enabling farmers to readily discern patterns, trends, and potential areas for enhancement.

REGULATORY FRAMEWORK AND COMPLIANCE REQUIREMENTS FOR AGRICULTURAL DRONE OPERATIONS

Drones or Unmanned Aircrafts essentially are small and agile machines flying in the air autonomously or remotely without having any pilot on board. The evolution of drone technology can be considered to be a significant technological revolution all over the world.

Drones have multi-dimensional utility in agriculture, mining, infrastructure, surveillance, emergency response, transportation, geo-spatial mapping, defence, law enforcement, etc. due to their versatility, ease of use, and reach, especially in India's remote and inaccessible areas. The uses of drones are extensively varied – from capturing aerial images to mapping difficult terrains for delivering goods, spraying pesticides over vast area of land, and to carry out surveillance in sensitive areas, etc., drones play an important role in providing access to areas which are usually inaccessible or are difficult to access.

The annual sales turnover of the drone manufacturing industry is expected to grow from INR 60 crore (roughly $8 million) in 2020-21 to over INR 900 crore (roughly $122 million) in FY 2023-24. The government has also come up with a profit linked incentive (PLI) scheme to give further impetus to drone and drone components manufacturing.

India, while recognising the importance of drones, enacted several rules to govern this area, with the most recent being the Drone Rules, 2021 notified by the Ministry of Civil Aviation on August 25, 2021. These new rules would replace the existing Unmanned Aircraft System (UAS) Rules, notified in March, 2021. The Drone Rules, 2021 are based on "trust, self-certification and nonintrusive monitoring", which will likely drive more investment and will promote innovation in the drone sector, thereby giving a leg-up to the drone industry.

The Aircraft Act, 1934 is the parent legislation that is in place with regards to the control of the manufacture, possession, use, operation, sale, import and export of aircrafts, which include the unmanned aircraft system, or in layman's words, the drone. The government also framed the Aircraft Rules, 1937 in terms of the mandate of the Act to provide for the detailed regulatory framework for aircrafts operating within the territory of India. The Aircraft

Rules have stringent provisions regarding the registration and operation of aircraft, which is justified given the higher risk involved in the operation of aircrafts. However, these stringent provisions, if applied to the drones or unmanned aircrafts, would have a detrimental effect as they create a barrier for innovations in the sector.

The pre-independence legislation and the rules, though amended from time to time, failed to recognize drones separately due to which, there was an elongated period of ambiguity and confusion related to governance of drones. Also, the remotely manned drones are substantially different than the manned aircrafts and therefore, an evolved and more comprehensive legal framework was required for the regulation of the drone sector.

In August 2018, the Director General for Civil Aviation (DGCA) for the first time issued the detailed requirements and compliance procedures under Civil Aviation Requirements (CAR) for remotely piloted aircraft system. However, these CAR rules were not elaborate and certain important aspects relating to national security were not addressed. Moreover, the requirements were restrictive for the new entrants and special exemptions would be required for most of the use cases, which made it difficult for stakeholders to undertake research and testing of drones.

Accordingly, the Ministry of Civil Aviation notified the Unmanned Aircraft System Rules, 2021 (the "UAS Rules") However, these Rules were narrowly focused on addressing the national security concerns by toughening the licenses and approval norms, which made them highly restrictive for other people. Hence, these Rules garnered a lot of criticism from people in academia, start-ups, endusers and other stakeholders stating to be restrictive in nature as they involved considerable paperwork, required permissions for every drone flight and very few "free to fly" green zones were made available.

To address the concerns of various stakeholders, the government floated draft rules in July 2021 and sought objections and suggestion from all the stakeholders. After taking into consideration the objections and suggestions, the Ministry of Civil Aviation notified the Drone Rules, 2021 in exercise of the powers conferred by the Aircraft Act, 1934. These Rules superseded the UAS Rules.

Drone Rules, 2021

Through these Rules, the government has considerably liberalised the regime on drones, by removal of express restrictions on foreign-owned and controlled Indian companies, by simplifying the drone registration and certification process and making it time-bound and less expensive, by creating an online platform (digital sky platform) for approvals and airspace map, by imposing lesser restrictions on nano drones, etc. However, these rules failed to deal with the privacy and safety concerns arising out of drone operations in open skies. This report discusses in detail the regulatory framework in place for drones in India.

Saliant Features of Drone Rules 2021

Built on a premise of trust, self-certification and non-intrusive monitoring.

1. Designed to usher in an era of super-normal growth while balancing safety and security considerations.

2. Several approvals abolished: unique authorisation number, unique prototype identification number,certificate of manufacturing and airworthiness, certificate of conformance, certificate of maintenance, import clearance, acceptance of existing drones, operator permit, authorisation of R&D organisation, student remote pilot licence, remote pilot instructor authorisation, drone port authorisation etc.
3. Number of forms reduced from 25 to 5.
4. Types of fee reduced from 72 to 4.
5. Quantum of fee reduced to nominal levels and delinked with size of drone. For instance, the fee for a remote pilot license fee has been reduced from INR 3000 (for large drone) to INR 100 for all categories of drones; and is valid for 10 years.
6. Digital sky platform shall be developed as a user-friendly single-window system. There will be minimal human interface and most permissions will be self-generated.
7. Interactive airspace map with green, yellow and red zones shall be displayed on the digital sky platform within 30 days of publication of these rules.
8. No permission required for operating drones in green zones. Green zone means the airspace upto a vertical distance of 400 feet or 120 metre that has not been designated as a red zone or yellow zone in the airspace map; and the airspace upto a vertical distance of 200 feet or 60 metre above the area located between a lateral distance of 8 and 12 kilometre from the perimeter of an operational airport.
9. Yellow zone reduced from 45 km to 12 km from the airport perimeter.
10. No remote pilot licence required for micro drones (for non-commercial use) and nano drones.
11. No requirement for security clearance before issuance of any registration or licence.
12. No requirement of Type Certificate, unique identification numberand remote pilot licence by R&D entities operating drones in own or rented premises, located in a green zone.
13. No restriction on foreign ownership in Indian drone companies.
14. Import of drones to be regulated by DGFT.
15. Requirement of import clearance from DGCA abolished.
16. Coverage of drones under Drone Rules, 2021 increased from 300 kg to 500 kg. This will cover drone taxis also.
17. DGCA shall prescribe drone training requirements, oversee drone schools and provide pilot licences online.
18. Remote pilot licence to be issued by DGCA within 15 days of pilot receiving the remote pilot certificate from the authorised drone school through the digital sky platform.
19. Testing of drones for issuance of Type Certificate to be carried out by Quality Council of India or authorised testing entities.

20. Type Certificate required only when a drone is to be operated in India. Importing and manufacturing drones purely for exportsare exempt from type certification and unique identification number.
21. Nano and model drones (made for research or recreation purposes) are exempt from type certification.
22. Manufacturers and importers may generate their drones' unique identification number on the digital sky platform through the self-certification route.
23. Easier process specified for transfer and deregistration of drones through the digital sky platform.
24. Drones present in India on or before 30 Nov 2021 will be issued a unique identification number through the digital sky platform provided, they have a DAN, a GST-paid invoice and are part of the list of DGCA-approved drones.
25. Standard operating procedures (SOP) and training procedure manuals (TPM) will be prescribed by DGCA on the digital sky platform for self-monitoring by users. No approvals required unless there is a significant departure from the prescribed procedures.
26. Maximum penalty for violations reduced to INR 1 lakh.
27. Safety and security features like 'No permission – no takeoff' (NPNT), real-time tracking beacon, geo-fencing etc. to be notified in future. A six-month lead time will be provided to the industry for compliance.
28. Drone corridors will be developed for cargo deliveries.
29. Drone promotion council to be set up by Government with participation from academia, startups and other stakeholders to facilitate a growth-oriented regulatory regime.

The DGCA's Regulatory Framework

The DGCA laid down a comprehensive set of guidelines in October 2014 to govern the use of drones in India. These regulations cover a wide array of aspects, from securing permissions to ensuring safety and privacy. It's crucial for drone operators in the agricultural sector to familiarize themselves with these rules to operate within legal boundaries and harness the full potential of drone technology.

Drone Classification and Registration

Drones are categorized based on their weight and intended use, influencing their registration requirements and operational permissions. Nano drones, the smallest category, weigh less than 250 grams and enjoy more freedom with minimal registration needs. Micro drones, weighing between 250 grams and 2 kilograms, also benefit from relaxed regulations for certain operations. However, for drones classified as small (2 to 25 kilograms), the requirements become more stringent, necessitating formal registration with the DGCA and specific approvals for flight operations.

Operational Limitations and Permissions

The DGCA's guidelines specify restricted zones and operational constraints to ensure public safety and national security. Areas around airports, military bases, and other sensitive locations are off-limits for drone activities. Additionally, special permissions are required for flights in controlled airspace or during nocturnal hours. Adhering to these guidelines is essential to prevent legal repercussions and ensure the safety of drone operations.

Data Privacy and Security

With the capability to capture detailed imagery and data, drones bring forth concerns regarding privacy and data security. Operators are mandated to implement measures, possibly including encryption, to safeguard the information collected during their flights. This is particularly significant in agriculture, where drones collect sensitive data about farms, crops, and livestock.

Insurance and Liability

Given the potential risks associated with drone flights, including accidents and damage to property or individuals, obtaining insurance is advisable for drone operators. Insurance coverage can mitigate financial liabilities arising from unforeseen incidents, ensuring that both the operator and affected parties are protected.

Adhering to Agricultural Laws and Standards

Drone use in agriculture is not only governed by aviation regulations but also by agricultural laws. These encompass safe pesticide application, environmental protection, and adherence to land use policies. Ensuring compliance with these regulations is vital for sustainable and lawful drone operations in the agricultural sector.

The Importance of Staying Informed

The regulatory landscape for drones is dynamic, with frequent updates and changes. For drone operators, staying informed about the latest regulations and technological advancements is crucial. This ensures that their operations remain compliant and take advantage of the latest innovations for increased efficiency and productivity.

The DGCA Rules and Regulations Regarding the Use of Drones in Agriculture

General Regulations

- **Registration and Licensing:** The DGCA requires registration for all drones, with the exception of nano drones (less than 250 grams) and non-commercial micro drones (less than 2 kg). Unless they are in a green zone and meet certain requirements, operators need a remote pilot license (RPL).
- **Green Zone:** The majority of agricultural areas are in the Green Zone, which exempts registered drones from flight restrictions below 120 meters in altitude. However, for up-to-date airspace information, visit the "DigitalSky" portal.

- **Guidelines for Safety:** Pre-flight checks, keeping to safe altitudes (usually 400 feet), steering clear of people and populated areas, and visual line-of-sight operation are all required.

Specific Agricultural Regulations

- **Training:** Completing a "agri-module" on agrochemical handling, mission protocols, and crop protection guidelines is a prerequisite for DGCA-certified pilots.
- **Permits:** Although general flight permits are not required in green zones, certain permits might be needed for certain activities such as chemical spraying or using larger drones. See your local government for more information.
- **Data privacy:** Information gathered from agricultural drone operations should be handled sensibly and in compliance with applicable laws.

Keep in mind that these are just general recommendations; location- and drone-specific laws may differ. Prior to using a drone for agricultural operations, make sure you are fully informed by consulting the most recent DGCA regulations as well as local authorities.

SIGNIFICANCE UNDERSTANDING DRONE LAWS IN INDIA

Understanding and adhering to drone regulations in India is paramount for individuals and businesses harnessing the power of unmanned aerial vehicles (UAVs), especially in sectors like agriculture.

The importance of comprehending and following these regulations can be highlighted through several key points:

- **Safety First:** Drone regulations are designed to ensure the safety of both airspace and ground activities. Adhering to guidelines prevents accidents, collisions, and potential damage, safeguarding people and property.
- **Airspace Management:** Drones share airspace with manned aircraft, and clear regulations help manage this shared space effectively. Understanding where and when drones can fly minimizes the risk of interference with commercial aviation, promoting overall airspace safety.
- **Privacy Protection:** Regulations outline boundaries that safeguard individuals' privacy. Drones equipped with cameras can capture images and videos, and compliance with regulations ensures that such data collection respects privacy rights and legal restrictions.
- **National Security:** Drones have the potential to be misused, posing a threat to national security. Regulations help control and monitor the deployment of drones, preventing unauthorized activities and potential security breaches.
- **Environmental Impact:** Drone usage, especially in agriculture, may involve the spraying of pesticides or fertilizers. Regulations guide responsible and sustainable practices, ensuring minimal environmental impact and promoting eco-friendly drone operations.
- **Data Security:** Drones often collect valuable data for various purposes. Regulatory compliance includes guidelines on data storage, sharing, and

protection, ensuring the security of sensitive information and preventing unauthorized access.

- **Insurance Requirements:** Regulatory frameworks often necessitate insurance coverage for drone operations. Understanding and meeting these requirements mitigate financial risks in the event of accidents, providing a safety net for operators.
- **Professionalism and Accountability:** Adhering to regulations demonstrates professionalism and accountability. It enhances the credibility of drone operators, whether they are farmers using drones for agriculture or businesses employing UAVs for various applications.
- **Legal Consequences:** Non-compliance with drone regulations can result in legal consequences, including fines and penalties. Understanding and following the rules is essential to avoid legal complications that could hinder the progress of drone-related activities.
- **Technology Advancement:** As technology evolves, regulations may be updated to accommodate new features and capabilities. Staying informed about these changes ensures that drone operators can leverage the latest advancements within the bounds of the law.

In conclusion, understanding and complying with drone regulations in India are not just legal obligations but essential steps toward fostering a safe, responsible, and sustainable integration of drones into diverse sectors, including agriculture. It ensures that the benefits of drone technology are realized without compromising safety, privacy, or national security.

Exceptions to Licensing Requirements

While obtaining a license for drone operations is a crucial aspect of compliance with India's regulatory framework, certain exemptions provide flexibility for specific scenarios. Understanding these exemptions is essential for drone operators looking to navigate the regulatory landscape efficiently.

- **Nano Drones:** Drones categorized as "nano" and weighing less than 250 grams fall under an exemption from the licensing requirement. These lightweight drones are considered less risky and are subject to fewer restrictions, making them accessible for hobbyists and casual users.
- **Micro Drones for Non-Commercial Use:** Microdrones, weighing between 250 grams and 2 kilograms, are exempt from licensing when used for non-commercial purposes. This exemption allows enthusiasts and hobbyists to engage in recreational drone activities without the need for a formal license.
- **Indigenous Manufacturers:** The DGCA provides exemptions for drones manufactured within India for research and development purposes. Indigenous manufacturers can test and showcase their products without the burden of obtaining a license, fostering innovation in the drone industry.
- **Government Agencies:** Drones operated by government agencies for specific activities, such as law enforcement, defence, or disaster management, may be exempt from licensing requirements. This exemption ensures that crucial public services can leverage drone technology swiftly in times of need.

- **Security-Sensitive Installations:** Drones operated within the vicinity of security-sensitive installations, such as airports, defence establishments, and strategic locations, may require specific approvals but could be exempt from standard licensing procedures. This is to prevent potential security threats while enabling necessary surveillance and monitoring.
- **Educational and Research Institutions:** Drones used for educational or research purposes by recognized institutions may receive exemptions from licensing requirements. This encourages the integration of drones into academic curricula and research projects without imposing unnecessary regulatory barriers.

Understanding these exemptions is vital for drone operators to leverage the benefits of unmanned aerial technology effectively. However, it's crucial to note that even when exempt, operators must comply with operational guidelines and safety measures outlined by the DGCA. Staying informed about the evolving regulatory landscape ensures that drone enthusiasts, researchers, and industry professionals can harness the potential of drones responsibly and in alignment with national safety and security standards.

Learning the Importance of DGCA Certification

Regulations and Compliance

Prior to acquiring an agricultural drone, it is imperative to verify its compliance with the regulations set forth by the Directorate General of Civil Aviation (DGCA). The DGCA establishes the criteria for drone operations in India, and compliance with these regulations is essential for lawful and secure drone utilization.

Benefits of Certification

DGCA certification beyond mere legal obligation; it signifies a standard of quality and safety. Certified drones adhere to the rigorous criteria established by aviation authorities, ensuring reassurance and mitigating potential legal issues.

Assesing Build Quality for Consistent Performance

Material Considerations: Metal versus Plastic

Build quality is a crucial consideration when choosing a drone. Regardless of whether you select a drone constructed from metal or plastic, sturdiness is essential for withstanding the demands of agricultural tasks, including rocky landings and severe environmental conditions.

Weatherproofing Attributes

Agricultural labor frequently entails erratic meteorological conditions. Drones with waterproof capabilities can function dependably in diverse settings, guaranteeing uniform data collecting irrespective of the climate.

Maximizing Flight Time for Optimal Coverage

Battery Capacity and Charging Duration

The duration of flight is a crucial factor in drone performance. Evaluate the battery capacity and charging duration to guarantee the drone can traverse large areas without recurrent disruptions.

Motor and Propeller Efficiency

The efficacy of a drone's motors and propellers directly influences flying duration. Select a drone equipped with sophisticated propulsion technologies that provide extended and more efficient flight durations.

Role of Megapixels in Agricultural Imaging

Advanced Imaging Technology for Precision Agricultural Surveillance

Megapixels are significant in agricultural imaging. High-resolution cameras yield detailed images crucial for precise crop monitoring and informed decision-making.

Reconciling Image Quality with Data Management

Although elevated megapixels enhance image quality, they concurrently escalate data storage and processing demands. Consider a drone that achieves an equilibrium between image quality and data size efficiency.

Planning Expenses for Accessories and Maintenance

Crucial Accessories for Enhanced Functionality

To enhance the utility of your drone, get essential extras such as more batteries and superior cameras. These enhancements can substantially improve your agricultural operations.

Evaluating Maintenance Expenditures

Consistent maintenance is essential for maintaining your drone in peak condition. Allocating funds for these expenses guarantees the sustained functionality and dependability of your drone over time.

Multi-Utility Drones: Versatility Matters

A multi-utility drone is capable of performing diverse functions including crop spraying, mapping, and monitoring, rendering it an adaptable instrument for agricultural use. This not only diminishes the necessity for numerous machines but also optimizes operations, conserving time and resources.

Loan and Subsidy Support for Farmers

Government financial support can facilitate farmers' investment in drone technology. Access to loans and subsidies can enhance adoption rates and propel breakthroughs in farming methods.

Choosing a Registered and Reputable Company

Acquiring goods from a licensed and esteemed organization guarantees the receipt of superior products and dependable customer service. This decision safeguards your investment and offers reassurance, ensuring that assistance is immediately accessible.

After-Sales Service and Support

Dependable post-purchase support is crucial for optimizing your drone's longevity and functionality. Consistent software upgrades and professional assistance can improve your drone's performance and maintain its alignment with the latest technical innovations.

Importance of Drone Training

Comprehensive training is essential for the safe and efficient use of drones. Training enables you to operate the drone, get precise data, and maintain the equipment, maximizing your investment's value.

The Value of a Warranty

A warranty signifies the manufacturer's assurance in their goods and guarantees that any problems will be rectified at no extra charge. This safeguard is essential for enduring satisfaction and confidence in your acquisition.

The Advantage of an All-India Dealer Network

An extensive network of dealers and distributors facilitates convenient access to maintenance and repairs, hence reducing downtime and maintaining your drone in optimal condition. The support of this network is essential for the sustained success of your agricultural drone investment.

2

Understanding the Components of a Drone

A drone is a sophisticated piece of technology made of many key components that work together to enable flight and various functions. The primary parts include the frame, which provides the structure and support for all other elements; the motors and propellers, which generate lift and control movement; the flight controller, acting as the brain of the drone, processing input from sensors and guiding the drone's flight path; and the battery, which powers the system. Moreover, sensors like GPS, accelerometers, and gyroscopes are integral for stability and navigation. Many drones also feature cameras and other payloads for specific tasks, such as aerial photography, surveying, or surveillance. These components, working in harmony, allow the drone to perform a wide range of applications in industries like entertainment, agriculture, and delivery.

KEY COMPONENTS OF A DRONE

Agricultural drones consist of multiple components, including a remote controller, an aerial platform, and a spraying mechanism. They possess sensors, cameras, and more components to enhance crop management and productivity.

Understanding the fundamental components is vital to understanding the potential of agricultural drones. This section delineates the essential components that render these drones viable instruments for contemporary agriculture.

Frame

The frame of an agricultural drone functions as its structural foundation. It is generally constructed from lightweight materials like carbon fiber or aluminum to maintain the drone's agility and aerodynamics while assuring longevity. The frame encompasses all other components and furnishes the essential structure for affixing the motors, propellers, and additional hardware.

Figure: 2.1 An Image Showing Frame of a Drone

The frame functions as the structural foundation of a drone, providing protection for all the delicate electrical components. It safeguards the internal electronics of the drone and furnishes mounting locations for diverse components, including FPV cameras, flight controllers, and video transmitters. The design, material, and dimensions of the frame significantly influence the drone's performance, stability, and flight characteristics.

Variants of Drone Frames

Drone frames are available in multiple configurations, each offering distinct advantages:

- **X-Frame:** This represents the predominant design. It provides excellent stability and agility, rendering it appropriate for racing and freestyle flight.
- **H-Frame:** This configuration offers increased capacity for components and can support larger batteries. It is frequently utilized for long-distance flights and accommodates greater payloads.
- **Hybrid X:** This integrates components of both X and H frames, providing a compromise between maneuverability and load-bearing capability.
- **Stretched X:** This is an extended variant of the X-frame, enhancing forward flying attributes and stability at elevated velocities.
- **Square Frame:** This configuration provides optimal internal capacity and is frequently utilized for photography drones or those equipped with numerous sensors.

Each frame type possesses distinct advantages, and the selection is contingent upon the drone's intended application, be it racing, freestyle flying, long-range missions, or payload transportation.

Choosing the Right Frame Material

The selection of frame material substantially influences the drone's weight, durability, and performance. The predominant materials utilized for drone frames are:

1. **Carbon Fiber:** This material is the preferred option for drone frames because to its remarkable strength-to-weight ratio. Carbon fiber frames are lightweight, robust, and inflexible, rendering them optimal for racing and freestyle drones. Nonetheless, they may be costly and challenging to manage.
2. **Aluminum Alloys:** Alloys like 6061 and 7075 offer an optimal combination of strength, lightweight properties, and cost-effectiveness. They are more manageable than carbon fiber, although less rigid.
3. **Fiberglass Composites:** These materials are occasionally utilized for cost-effective or amateur models. They provide a lightweight and robust alternative, however are less prevalent than carbon fiber or aluminum.
4. **High-Density Polyethylene (HDPE):** Although not as robust as carbon fiber, HDPE frames exhibit more flexibility and resilience, potentially enhancing overall durability.

Certain frames include a composite of materials, including carbon fiber with aluminum or 3D printed TPU, to enhance strength, safeguard components, or augment the design's aesthetic appeal.

Factors to Consider when Choosing Frame Size

Frame size is generally assessed by the diagonal measurement between motors and is frequently classified according to the propeller size it can support. This is a comprehensive guide on frame dimensions and their applications:

Table 2.1 Frame Dimensions and Their Uses

Frame Size	Prop Size	Common Use
280mm+	7 inch	Long Range
250mm	6 inch	Long Range / Freestyle / Racing
210mm	5 inch	Freestyle / Racing
180mm	4 inch	Freestyle / Racing
150mm	3 inch	Freestyle / Racing
112mm	2 inch	Freestyle / Racing
<100mm	Whoops	Indoor / Limited outdoor

The predominant frame size is around 210mm, commonly designated as a 5-inch frame. This size provides the optimal balance of agility, weight, and response, rendering it appropriate for both freestyle and racing.

When selecting a frame size, take into account elements such as portability, maneuverability, flying duration, and cargo capacity. Larger frames can support larger batteries for extended flight durations and transport heavier payloads, although smaller frames are generally more nimble and reactive.

The frame size dictates the dimensions of the propellers and motors permissible, thereby influencing the drone's performance and capabilities. Selecting a frame size that corresponds with your planned application, be it racing, freestyle flying, or aerial photography, is essential.

Motors

Motors serve as the driving force behind agricultural drones. These electric motors produce the requisite thrust to elevate the drone from the ground and maintain its flight. Agricultural drones commonly employ brushless DC motors due to their efficiency and dependability. The quantity and dimensions of motors may differ based on the drone's configuration and load capacity.

Figure: 2.2 Motor

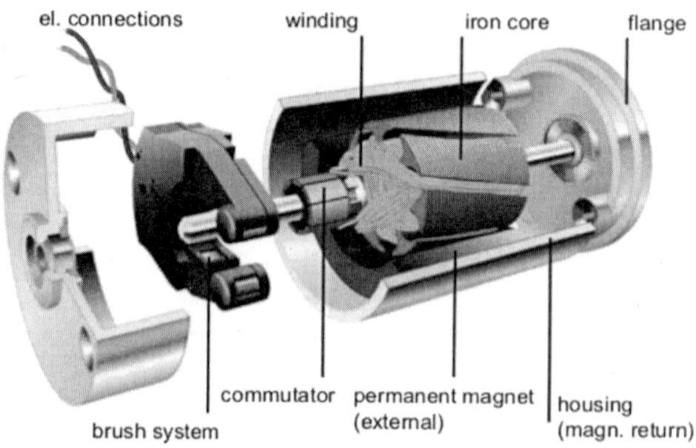

Figure: 2.3 Brushless Vs Brushed DC Motors

Motors are the core components of every drone, responsible for producing the thrust required for flight. They are essential in assessing a drone's performance, efficiency, and overall capabilities. Comprehending the various engine types and their attributes is crucial for both drone aficionados and experts.

Comparison of Brushed and Brushless Motors

Drones often employ two categories of motors: brushed and brushless. Although both function based on electromagnetism, they exhibit substantial differences in design and performance attributes.

Brushed motors use a revolving stator (wire coils) that functions as an electromagnet with dual poles. The shell remains fixed, facilitating effortless installation without screws in several instances. These motors are typically more cost-effective to manufacture and may be produced in compact sizes, rendering them appropriate for nano drones and miniature toys.

Brushless motors utilize a permanent magnet and electrically alternate polarity rather than employing brushes. This design presents multiple benefits:

1. Enhanced efficiency (85-90% in contrast to 75-80% for brushed motors)
2. Extended longevity resulting from the elimination of brushes that deteriorate.
3. Enhanced thermal dissipation and broader velocity spectrum
4. Enhanced velocity and torque, particularly at elevated speeds

Consequently, brushless motors have emerged as the favored option for the majority of FPV drones, ranging from micro-sized variants to substantial heavy-lift versions.

Motor Size and KV Rating

The size of motors in drones is generally represented by a four-digit numeral, for example, 2207 or 2306. The initial two figures denote the stator diameter in millimeters, whilst the final two signify the stator height. A 2207 motor possesses a stator width of 22 mm and a height of 7 mm.

The dimensions of the motor influence its performance attributes:

- Larger motors typically generate greater thrust but are also heavier.
- Taller stators often provide increased top speeds but may exhibit inferior low-speed handling.
- Wider stators enhance low-speed maneuverability but may result in reduced maximum velocities.
- The KV rating is a vital consideration in motor choosing. It denotes the theoretical revolutions per minute (RPM) the motor will achieve per volt of power in the absence of a load. A 2300KV motor fueled by a 4S (14.8V) battery would potentially achieve 34,040 RPM without propellers.
- The selection of KV rating is contingent upon the intended application and battery configuration:
- Higher KV motors are typically more appropriate for racing and high-speed applications.
- Lower KV motors provide increased torque and are frequently utilized with larger propellers to enhance efficiency and extend flying durations.

Motor Mounting and Wiring

- Correct motor installation is essential for drone efficacy and safety. Contemporary 5-inch quadcopters predominantly utilize motors sized between 2205 and 2407, generally exhibiting mounting patterns of 16x16mm or 16x19mm. During the installation of motors:
  1. Verify that all four mounting screws are utilized and fastened securely.
  2. Place the motor on the arm, rather than beneath it.
  3. Observe the directional indicators on the motor shaft, as they denote the appropriate rotation direction.
- Brushless motors necessitate a three-wire connection to the electronic speed controller (ESC), whereas brushed motors require only two wires. When connecting brushless motors:
  1. Thread the three wires from the motor through the specified aperture in the arm.
  2. Attach the wires to the ESC, ensuring the correct sequence for the intended rotation direction.
  3. Employ cable management strategies to maintain wire organization and avert interference with propellers.

By comprehending these fundamental elements of drone motors, amateurs and professionals can make informed choices when constructing or enhancing their drones, so ensuring optimal performance for their particular requirements and applications.

Propellers

Propellers generate the thrust necessary for elevating and propelling the drone. Agricultural drones frequently utilize specifically engineered propellers intended for efficiency and stability in airborne activities. The dimensions, pitch, and rotational orientation of the propeller are meticulously chosen to align with the drone's design and purpose.

Figure: 2.4 Propellers

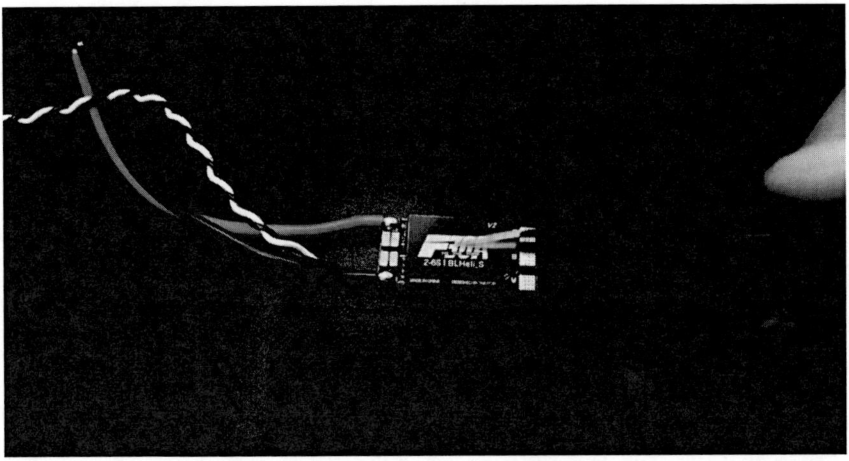

Figure: 2.5 Model of a Propeller

Propellers are critical elements of FPV drones, influencing their power, smoothness, and response. These customized airfoils are affixed to the motor hub and are available in diverse forms, sizes, and blade quantities. Comprehending propellers is essential for enhancing a quadcopter's performance.

Drone propellers function by rotating at elevated velocities to provide thrust, so generating lift and advancing the drone forward. The rotating blades produce a disparity in air pressure, resulting in lower pressure at the blade's front compared to the back, so providing thrust to propel the drone.

Propeller Size and Pitch

- Propeller dimensions are often expressed in imperial inches, with one inch equivalent to 2.54 centimeters. The specifications for propellers encompass length, pitch, and blade count. A 6x4.5x2 propeller (designated as 6045x2) measures 6 inches in length, features two blades, and possesses a pitch of 4.5 inches.
- Propeller pitch denotes the distance a propeller advances in one complete revolution, quantified in inches. It essentially denotes the distance the propeller would advance if traversing a solid medium rather than air. A propeller with a higher pitch displaces more air per revolution, producing greater thrust at elevated speeds while yielding diminished thrust while the aircraft is stationary.
- The required propeller size is contingent upon the drone's dimensions, mass, and particular application specifications. Larger drones demand larger propellers to generate adequate thrust, but smaller drones can utilize smaller propellers.

Material Options for Propellers

Drone propellers are generally constructed from plastic, carbon fiber, or wood, each possessing distinct attributes:

1. **Plastic propellers:** are the most prevalent type utilized in numerous home-built quadcopters. They are lightweight, cost-effective to produce and acquire, and simple to balance.
2. **Carbon fiber propellers:** While more costly than plastic alternatives, carbon fiber propellers are exceptionally lightweight, facilitating the use of larger propellers with rapid acceleration and deceleration times. They possess remarkable strength, which can be advantageous but may transmit greater force to the motor and airframe in the event of a collision.
3. **Wooden propellers:** Although prevalent in model airplanes, wooden propellers are infrequently utilized in multi-rotors due to their weight. They are more difficult to accelerate or decelerate than plastic or carbon fiber propellers.

Electronic Speed Control (ESC)

The Electronic Speed Control (ESC) for drones is an essential electronic component that regulates the velocity and orientation of each motor within the drone. It reads data from the flight controller, modulating the power delivered to the motors to regulate their revolutions per minute (RPM) and, hence, the drone's movement and stability in flight. It serves as a conduit between the flight controller and the motors, guaranteeing accurate control and maneuverability of the drone.

Figure: 2.6 Electronic Speed Control (ESC)

Understanding ESC Ratings

- When choosing an ESC for your drone construction, it is crucial to evaluate two primary specifications: current and voltage. The present rating, expressed in amperes, denotes the maximum current an ESC can endure without incurring harm. Electronic Speed Controllers (ESCs) often possess two current specifications: continuous and burst. The continuous rating denotes the stable current an ESC can reliably handle, but the burst rating signifies the peak current it can withstand for short durations, often less than 10 seconds.

- It is essential to select an ESC with a current rating that surpasses the maximum current draw of your engine and propeller configuration. An amp rating can be excessively high, but it can be insufficient, which may result in ESC failure during flight.
- The voltage rating is of equal significance. Electronic Speed Controllers (ESCs) must align with the voltage of your battery to avert damage. Contemporary ESCs accommodate input voltages ranging from 4S to 6S LiPo batteries, with certain models capable of handling even greater voltages.

Choosing the Right ESC for Your Build

When selecting an ESC for your drone build, consider the following factors:

1. Current and voltage ratings
2. ESC type: Single or 4-in-1
3. Firmware compatibility
4. ESC protocol support
5. Size and weight

Single ESCs are discrete units that regulate a single motor apiece. They provide superior cooling owing to their placement on the drone's arms and are more readily replaceable if compromised. 4-in-1 ESCs integrate four electronic speed controllers onto a single circuit board, facilitating installation and minimizing weight.

ESC protocols dictate the communication velocity between the flight controller and the ESC. Contemporary protocols such as DShot (DShot150, DShot300, DShot600) provide enhanced performance and are advised for utilization with Betaflight.

When selecting an ESC, verify its capability for the required protocol and its compatibility with your flight controller and motors. Furthermore, take into account the ESC's dimensions and mass, as these elements can influence your drone's overall performance and flight attributes.

By meticulously evaluating these parameters, you may choose the most suitable ESC for your FPV drone assembly, guaranteeing excellent performance and dependability during flight.

Flight Control Board

The flight control board, commonly known as the drone's "brain," processes input from many sensors and transmits commands to the motors and ESCs. It is essential for stabilizing the drone, maintaining its level orientation and precisely responding to pilot commands. Flight control boards utilize sophisticated algorithms and sensors, such as accelerometers and gyroscopes, to provide stability and accurate control.

The flight controller (FC) works as the drone's central nervous system, orchestrating its motions and operations. This compact circuit board, analogous to a computer's motherboard and CPU, contains sophisticated electronics and software that oversee and regulate all facets of the drone's functionality. The flight controller (FC) interfaces with multiple sensors to collect essential data regarding the drone's altitude, orientation, and velocity, utilizing components such an Inertial Measurement Unit (IMU), barometer, and distance sensors.

Figure: 2.7 Flight Controller

The principal role of a flight controller is to analyze sensor data and pilot commands to determine the requisite speed for each motor of the drone. It subsequently relays these directives to the Electronic Speed Controllers (ESCs), which convert the commands into signals comprehensible to the motors. This complex mechanism enables the drone to pivot and accelerate by generating speed variances among its motors.

Various Flight Controller used in Agriculture Drones

The flight controllers and modules specifically engineered for agricultural uses include:

- The JIYI K++ V2 Flight Controller is acclaimed for its accuracy. This flight controller is especially esteemed for its dependability in agricultural drone operations.
- The JIYI Ground Proof Terrain Follow Radar is crucial for sustaining a stable altitude over varied terrains. To guarantee precise application and oversight.
- The JIYI Ground Proof Terrain Follow Radar is specifically engineered for compatibility with the JIYI K++ V2 Flight Controller.
- JIYI Obstacle Avoidance Radar facilitates the detection and evasion of impediments during flight. This functionality is essential for secure operations.
- JIYI Obstacle Avoidance Radar for agricultural drones is designed to operate in intricate situations.
- SIYI DK32 Air Unit 2.4G Receiver with Telemetry Control is compatible with the DK32 Remote Controller, ensuring dependable connection and control. This is crucial for extended-range activities.
- JIYI CAN Hub-12S Module improves connectivity and integration of various components.

- The JIYI CAN Hub-12S Module for K++ is fully compatible with the JIYI K++ flight controller system for agricultural drones. This improves total functionality.

Popular Flight Controller Options

Flight controllers are available in diverse varieties, addressing distinct user requirements and proficiency levels:

1. **Hobbyist/Builder flight controllers:** These are simple to install and cost-effective, making them suitable for novices.
2. **Racing football cleats:** Engineered for lightweight, precise, and responsive performance, generally priced below €50.
3. **Filming FCs:** Concentrated on producing seamless shots and facilitating user-friendly operation, frequently incorporated into camera drones such as those manufactured by DJI.
4. **Commercial flight controllers:** Sophisticated devices for the transfer of high-value goods and secure operation, featuring prominent manufacturers such as DJI, Pixhawk, Auterion, and Fusion Engineering.

There are two primary categories of flight controllers for multirotor drones:

1. **Conventional Flight Controller:** Employs distinct boards for the flight controller and electronic speed controllers (ESCs).
2. **All-in-One (AIO) Flight Controller:** Consolidates the Electronic Speed Controller into a single board, providing a compact configuration while possibly compromising certain durability aspects.

Key Features of Flight Controller

When choosing a flight controller, take into account the following critical attributes:

1. **Processor:** Contemporary FCs utilize STM32 microcontrollers, including the F4, F7, and H7 models. The F7 and H7 processors provide exceptional performance for intensive activities, but the F4 serves as a budget-friendly alternative.
2. **Gyroscope:** The gyroscope is essential for equilibrium and responsiveness. Common selections are the MPU6000 and the more recent ICM series, which provide enhanced performance when soft-mounted.
3. Select a flight controller that accommodates a broad spectrum of battery configurations, preferably up to 6S, to facilitate experimenting.
4. **Vibration Dampening:** Seek controllers equipped with efficient dampening mechanisms, such as silicone grommets, to reduce undesirable vibrations.
5. Select controllers featuring dedicated connections for digital video systems and facilitating seamless interaction with VTX units, ESCs, and additional FPV components.
6. A barometer, though not obligatory, can improve the precision of GPS-assisted flight.

7. **OSD Chip:** For analog FPV systems, verify that the flight controller is equipped with an OSD chip (AT7456E) to enable Betaflight OSD capabilities.

Radio Receiver

The radio receiver acquires signals from the drone operator's transmitter, converting them into commands that the flight control board can implement. This component facilitates remote operation of the drone, encompassing capabilities such as throttle, pitch, roll, and yaw, so enabling the pilot to maneuver and direct the drone's motions.

Understanding Drone Transmitters for Effective Control

The radio transmitter, commonly referred to as the TX in drone terminology, is essential for operating your drone. There exist fundamental models that enable you to manage the essentials. Additionally, there exist more sophisticated models featuring LCD screens, programmable buttons, and telemetry displays. These devices are available in many sorts and functionalities.

The transmitter's range is crucial. It indicates the maximum distance for maintaining a reliable connection with your drone. The transmitter's communication method must align with the drone's receiver functionality to ensure seamless operation.

Additionally, distinct transmitters enable the management of several channels. This indicates your capacity to manage multiple functions concurrently. Understanding your transmitter's characteristics enhances your ability to control and appreciate the flight experience.

The Role of Receivers in Drone Operations

The drone's receiver is essential. It transmits your control inputs from the transmitter to the drone's flight controller. This electronic component receives radio signals from the transmitter and converts them into commands usable by the drone. For optimal performance, the receiver requires a properly fitted and oriented antenna. The antenna captures the incoming radio waves.

Various factors can influence the efficacy and range of communication. This encompasses the receiver's quality, the antenna type, and potential interference from external sources. Certain drones utilize a device that amalgamates the transmitter and receiver, referred to as a transceiver. This design is more compact and practical.

Maintaining a reliable connection with your drone is essential for optimal receiver functionality. This additionally aids in preventing signal degradation during flight. Regularly inspect the receiver's connections, select high-quality antennas, and remain vigilant regarding potential interference. These measures can facilitate a secure and enjoyable flying experience.

Sensors

Sensors form the sensory array of agricultural drones, each designed for specialized functions:

- **Cameras:** Agricultural drones frequently possess cameras that can capture high-resolution photos, which may encompass RGB (visible light), multispectral, or thermal imaging sensors.
- **LiDAR (Light Detection and Ranging):** LiDAR sensors employ laser beams to generate intricate 3D representations of the environment, rendering them advantageous for topographical mapping and elevation analysis.
- **Infrared Sensors:** Infrared sensors are employed for assessing crop health, recognizing plant stress, and measuring temperature.

Ultrasonic sensors assist drones in sustaining a stable altitude above the ground, particularly during low-altitude operations.

Battery

The battery serves as the principal power source for the agricultural drone. Lithium-polymer (LiPo) batteries are frequently utilized because of their elevated energy density and lightweight characteristics. The battery's capacity and voltage directly affect the drone's flight duration and payload capability. Sophisticated battery management systems provide secure and efficient energy distribution.

GIMBALS AND CAMERAS

Agricultural drones frequently incorporate gimbals to stabilize the cameras and sensors affixed to them. A gimbal enables the camera or sensor to preserve a stable orientation, regardless of turbulence or directional shifts experienced by the drone. This consistency is essential for obtaining precise and applicable data during agricultural surveys and inspections.

Figure: 2.8 Drone with a Camera

Utilizing a gimbal and camera configuration is essential for achieving high-quality and stable aerial photography and videography. Gimbals function as steady platforms. They stabilize the drone camera against vibrations and movements.

This enables the capture of smooth, cinematic footage, even in windy conditions or during rapid movement.

The specifications and characteristics of your camera are significantly important. The type of camera, whether an action camera or an integrated drone camera, will influence the quality and experience of your video recording. Key attributes encompass resolution, frame rate, and sensor dimensions. Take these factors into account while choosing a camera.

Furthermore, sophisticated gimbals offer features such as object tracking, intelligent flight modes, and live streaming capabilities. Investing in a quality gimbal and camera configuration can significantly enhance your aerial photography and filming endeavors.

FUNCTIONALITY OF EACH COMPONENT AND ITS ROLE IN DRONE OPERATION

Table 2.2 Drone Parts and Their Functions

Components	Functions
Airframe	The structural backbone of a drone, the airframe, supports all components, including propulsion systems, payloads, and electronic systems. Designed with precision, it balances durability with lightness, enabling drones to withstand environmental challenges while maintaining manoeuvrability.
Propulsion System (Motors, Propellers, ESCs)	This system is the lifeline of a drone, dictating its thrust, lift, and overall flight dynamics. Motors and propellers are calibrated to work seamlessly with Electronic Speed Controllers (ESCs), providing the precise control needed for diverse operations, from aerial photography to precision agriculture.
Power Source	Typically, lithium-polymer (LiPo) batteries fuel ideaForge drones, offering an optimal blend of weight and energy efficiency. This ensures extended flight times and reliability, essential for missions requiring endurance and resilience.
Flight Controller	Considered the brain of the drone, the flight controller processes inputs from sensors and the remote control to adjust the drone's behaviour in real time. Its advanced algorithms allow for stable flight, autonomous navigation, and precise positioning, crucial for both novice and expert pilots.

Components	Functions
Sensors and Navigation Systems	From GPS modules that enable precise location tracking to gyroscopes and accelerometers that maintain stability and orientation, these sensors are integral for autonomous flight capabilities. Advanced UAVs may also include obstacle detection and avoidance systems, enhancing their safety and utility in complex environments.
Camera and Gimbal	High-quality imaging is a staple of ideaForge drones, with integrated cameras and gimbals offering stabilised, high-resolution photos and videos. These components are vital for surveillance, inspection, and mapping applications, where clarity and detail are paramount.
Communication System	Robust remote control and data transmission systems ensure seamless communication between the drone and its operator. This includes not only the control link but also video streaming capabilities, allowing for real-time monitoring and decision-making.
Software and Firmware	The unsung heroes of drone operations, these programs integrate hardware components, enabling complex flight patterns, data analysis, and user-friendly interfaces. They ensure that drones from ideaForge are not just tools, but partners in achieving operational goals.

The Role of Small Electronic Components in a Drone

While we have addressed the primary components of a drone, a comprehensive understanding of its operation necessitates an examination of the electronic elements. Specific electronic components serve distinct functions in a drone's capabilities.

- Crystal oscillators facilitate a drone's wireless communications, micro-processing at both ends, Bluetooth and/or Wi-Fi connectivity, timing capture, camera operation, sensor functionality, and navigation. Any drone equipped with autonomous navigation utilizes a diverse range of sensors, necessitating precise timing solutions.
- Crystal-controlled clock oscillators (XOs) are essential for communication systems and microprocessor timing. Inductors are employed to provide pure DC power, essential for data processing, video transmission, and communications.
- Temperature-compensated crystal oscillators (TCXO) are utilized for GPS and radio communications to and from the drone. These electrical components

are crucial for GPS functionality, as precise timing and signal accuracy are essential for determining longitude and latitude. Fundamentally, superior timing solutions enhance the efficacy of the GPS system.

Certain drones utilize global navigation satellite systems (GNSS) for navigation, necessitating precise timing solutions integrated into its SDR time board. The internal time board employs an oven-controlled crystal oscillator (OCXO) due to its provision of a stable and precise signal, little noise floor, low phase noise, and superior frequency stability. GNSS systems impose stringent timing requirements, making the OCXO timing solution from ECS Inc. the optimal choice for such drones.

WORKING OF MODERN DRONES

Aerodynamics and Lift

Rotary Wing Drones (Quadcopters)

- Rotary Wing Unmanned Aerial Vehicles (Quadcopters):

Most prevalent drones, such as quadcopters, utilize several rotors to produce lift.

Each Rotor Comprises a Motor and a Propeller.

- The rotation of the propellers generates a downward thrust by displacing air, which subsequently produces an upward lift force, enabling the drone to hover, climb, or plummet.

Fixed Wing Drones

- These drones function similarly to conventional aircraft, use fixed wings to produce lift.
- They necessitate forward propulsion, supplied by a propeller or jet engine, to produce airflow over the wings and generate lift.

Propulsion System

- **Motors:** Electric motors are generally employed to rotate the propellers. These motors operate on battery power.
- **Propellers:** The configuration and pitch of the propellers are essential for producing the requisite thrust to elevate the drone.

Control System

- **Flight Controller:** The primary processing unit of the drone that receives inputs from sensors and the pilot's directions to regulate the motors and ensure stable flight.
- **Sensor:** Drones are outfitted with an array of sensors, such as gyroscopes, accelerometers, magnetometers, GPS, barometers, and occasionally cameras or ultrasonic sensors. These sensors furnish information regarding the drone's orientation, velocity, altitude, and location.

- **Electronic Speed Controllers (ESCs):** These devices modulate motor speed according to directives from the flight controller.

Communication System

- **Remote Control:** Drones are generally operated with a remote control or a smartphone application that connects with the drone through radio waves, typically at frequencies of 2.4 GHz or 5.8 GHz.
- **Telemetry:** Certain drones transmit real-time data to the operator, encompassing battery status, GPS coordinates, and camera feed.

Power System

- **Battery:** Drones utilize rechargeable lithium-polymer (LiPo) batteries to supply power to the motors, sensors, and other electrical components.

Navigation and Positioning

GPS

- Numerous drones utilize GPS for navigation and positional stability.
- GPS facilitates accurate management of the drone's position and supports functionalities such as waypoint navigation and return-to-home capabilities.

Obstacle Detection and Avoidance

Advanced drones possess sensors that identify impediments in their trajectory and modify their flying path to prevent collisions.

Stabilization and Maneuvering

- **Stabilization Algorithms:** The flight controller employs algorithms to analyze sensor data and execute swift modifications to the motors, so assuring stable flying. This encompasses the maintenance of equilibrium (attitude control) and the regulation of yaw, pitch, and roll.
- **Maneuvering:** The drone can execute diverse movements, rotations, and intricate maneuvers by adjusting the speed of its individual motors.

Payload

- **Cameras and Sensors:** Numerous drones are outfitted with cameras or alternative sensors for diverse uses including photography, videography, surveying, and environmental monitoring.
- **Delivery Mechanisms:** Delivery drones are equipped with systems to transport and deploy packages.

OPERATING A DRONE

Operating a drone can be an enjoyable and fulfilling endeavor; nevertheless, it is crucial to understand how to do it securely and in accordance with legal regulations.

Below are the fundamental procedures for operating a drone:

1. **Consult the Manual:** The initial step is to thoroughly examine the drone manual, since each model possesses distinct instructions.
2. **Register your Drone:** Depending on your jurisdiction, it may be necessary to register your drone with the relevant authorities.
3. **Charge your Drone Battery:** Ensure your drone battery is completely charged prior to operation.
4. **Select an Appropriate Site:** Opt for a spot that is unobstructed, clear, and distanced from any impediments such as trees, structures, or electricity lines.
5. **Verify the Weather:** Refrain from using your drone in windy or inclement conditions. Consult the meteorological prediction before to flying.
6. **Turn on the Drone:** Activate the drone and the remote control.
7. **Calibrate the Drone:** Adhere to the manual's directions to calibrate the drone prior to flight.
8. **Initiate Ascent:** Gradually and steadily advance the throttle stick to elevate the drone.
9. **Descend the Drone:** When prepared to land, gradually lower it to the earth utilizing the throttle stick.
10. **Land the Drone:** Upon landing, switch off both the drone and the remote control.
11. **Deactivate the Drone:** Upon landing, switch off both the drone and the remote control.

Figure: 2.9 Common Equipments for Operating a Drone

Safety Measures During Drone Operation

Drones can be an enjoyable and beneficial instrument; nevertheless, they may pose risks if misused. Consider the following precautions when operating a drone:

Know the Laws and Regulations

- Prior to operating a drone, ensure you are familiar with the laws and regulations applicable in your vicinity.
- This encompasses all local, state, and federal regulations, along with any limitations on drone flight locations.

Always Keep your Drone in Sight

- It is essential to maintain visual contact with your drone at all times.
- This will assist you in evading collisions with other items or individuals.

Fly in Open Areas

- Attempt to operate your drone in unobstructed regions, distant from individuals, structures, and other impediments.
- This will assist in preventing accidents and collisions.

Respect People's Privacy

- Do not operate your drone over private property without obtaining authorization.
- Avoid operating your drone near residential properties or in locations where individuals possess a legitimate expectation of privacy.

Avoid Flying in Bad Weather

- Drones are not engineered to withstand severe weather conditions such as high winds, intense rainfall, or snowfall.
- Avoid operating your drone under these conditions, as they may result in a crash or damage to the device.

Keep Your Drone in Good Condition

- Consistently inspect your drone for indications of damage or deterioration.
- Replace any compromised components prior to flight.

Practice Safe Battery Use

- Utilize only the batteries and charger specified by the manufacturer.
- Refrain from charging batteries without supervision and do not utilize broken or bloated batteries.

Be Prepared for Emergencies

- **Prepare for Emergencies:** Maintain proximity to a first aid kit and a fire extinguisher for potential crises. Additionally, be ready to execute a rapid landing of your drone if required.

By adhering to these precautions, you can facilitate a secure and pleasurable drone experience for both yourself and others.

BASICS OF DRONE AERODYNAMICS AND FLIGHT PRINCIPLES

Working Principle of Drone and Flow Pattern

Fluid dynamics is crucial in the design and development of aircraft and drones. This topic encompasses the operational principles of aircraft aerodynamics.

- A requisite magnitude of upward force, referred to as Lift, is necessary to elevate the vehicle against gravitational pull.
- A force generated to propel a vehicle or body in motion is termed thrust. The kinematic rules of fluid flows can be employed to analyze these forces.

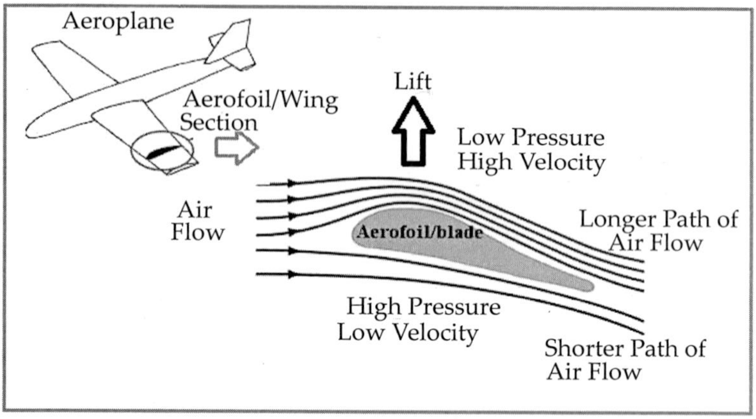

Figure: 2.10 Principle of Aerodynamic Lift

When air traverses an aerofoil, pressure, viscosity, and drag forces have influence on the profiles Force is exactly proportional to the inlet air velocity.

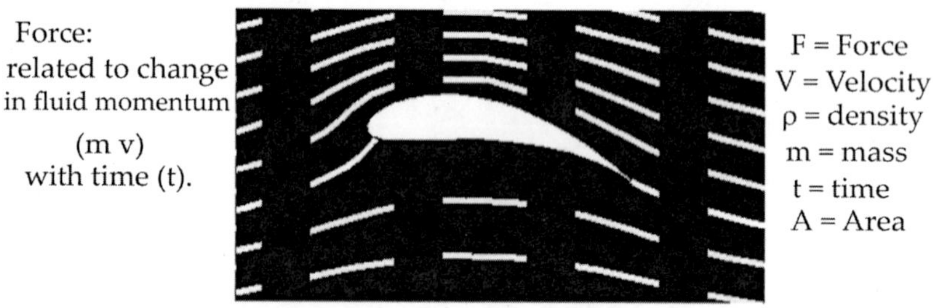

$$F = \text{Constant } (mV)/t$$

Re-write equation: $F = \text{Constant } (m/t)/V$

Use definition of mass flow rate (m/t): $m/t = \rho V A$

$$F = \text{Constant } (\rho VA)\ V$$

Change the constant to include A: $F = \text{Constant } (\rho V V)$

$$F = \text{Constant} \times \rho \times V^2$$

Double the Velocity $\rightarrow$ Quadruple the Force

Half the Density $\rightarrow$ Half the Force

Figure: 2.11 Profiles Force is Exactly Proportional to the Inlet Air Velocity

The flow configuration surrounding the cross-section of the aerofoil or propeller is illustrated below. Elevated fluid pressure at the base and diminished pressure at the apex of the propeller generate an upward force known as lift. This force is accountable for elevating the mass of an aircraft or unmanned aerial vehicle.

The magnitude of lift force is contingent upon the angle of inclination of the airfoil or propeller.

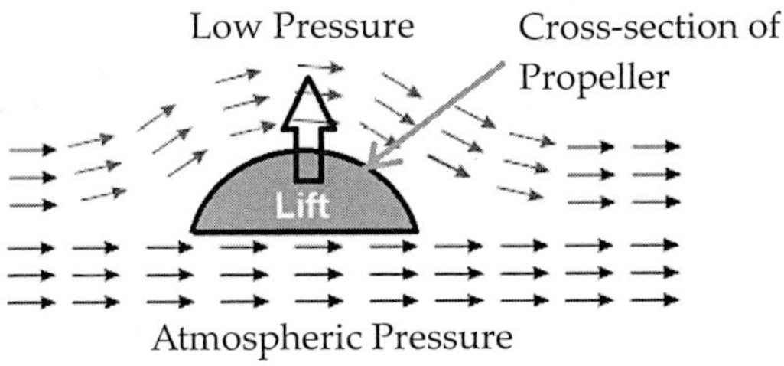

Figure: 2.12 The Flow Configuration Surrounding the Cross-section of the Propeller

According to the principle of conservation of energy in fluid dynamics (Bernoulli's principle), the total energy in a fluid remains constant along a streamline.

- When air traverses an aerofoil or wing, its velocity amplifies at the upper section. However, the atmospheric pressure diminishes.
- Conversely, the air velocity diminishes while pressure escalates on the underside of the blade. The subsequent pressure differential over the aerofoil generates an upward force known as lift.
- CFD simulation of flow across an airfoil is crucial in various automotive and aerospace industries.

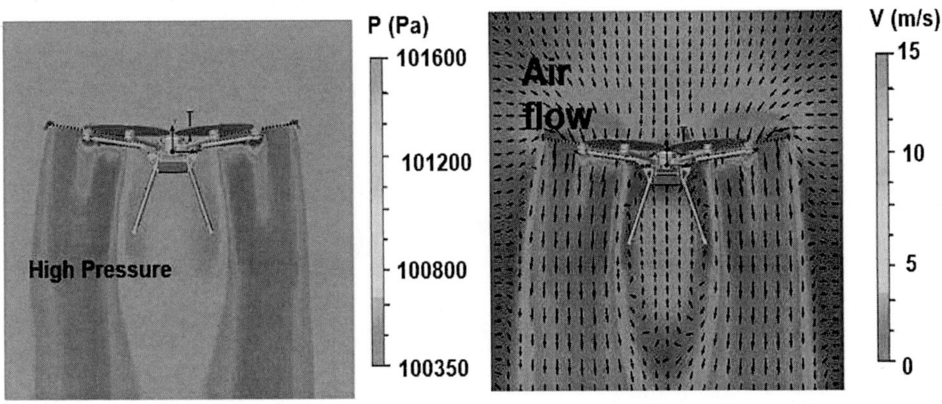

Figure: 2.13 Total Pressure Contours and Velocity Contours

Frones, also known as Unmanned Aerial Vehicles (UAVs), operate based on several key principles involving aerodynamics, propulsion, control systems, and onboard electronics.

WORKING PRINCIPLE OF QUADCOPTER

A quadcopter features four propellers positioned at the corners of its frame.

- Each propeller's speed and rotational direction are independently regulated to ensure the drone's balance and movement.
- In a conventional quadrotor, all four rotors are positioned equidistantly from one another.
- To preserve system equilibrium, one rotor pair rotates clockwise while the other pair rotates counterclockwise.
- All rotors must operate at high speed to achieve lift (hover). Altering the rotor speed enables the drone to maneuver forward, backward, and laterally.

Quadcopter Dynamics

- The movement of drones is categorized into four types depending on the motion relationship among the four propellers: 1) throttle, 2) pitch, 3) roll, and 4) yaw.
- The intricacies of quadcopter dynamics are elucidated in numerous literature.

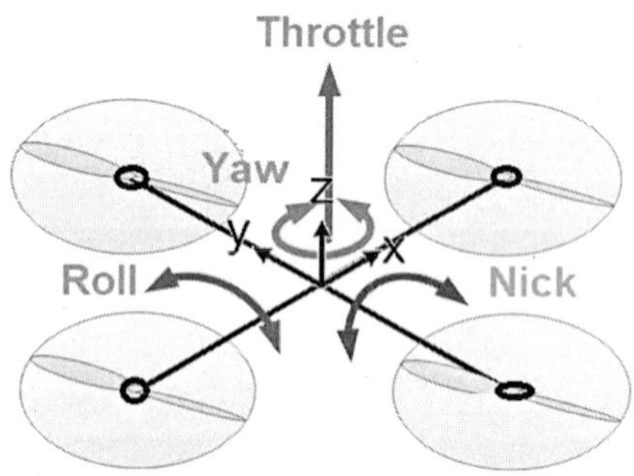

Figure: 2.14 Operational Mechanism of Drone-Quadcopter Dynamics

- **Throttle/ Hover:** The vertical movement of the drone is referred to as throttle or hover.
  - If all four propellers operate at standard velocity, the drone will descend.
  - If all four propellers operate at an increased velocity, the drone will ascend. This phenomenon is referred to as drone hovering.
- **Pitch:** The movement of a drone along a lateral axis, either forward or backward, is referred to as pitching motion.
  - If both rear propellers operate at high velocity, the drone will advance ahead.
  - If both front propellers operate at high velocity, the drone will advance in the reverse direction.
- **Roll:** The motion of a drone around its longitudinal axis is referred to as rolling motion.

- If two right propellers operate at high velocity, the drone will traverse to the left.
- If both left propellers operate at elevated speeds, the drone will go to the right.

- **Yawn:** The rotation of the drone's head around the vertical axis, either to the left or right, is referred to as yawing motion.
    - If two propellers of a right diagonal operate at high velocity, the drone will rotate in an anti-clockwise manner.
    - If two propellers on the left diagonal operate at high velocity, the drone will rotate clockwise.

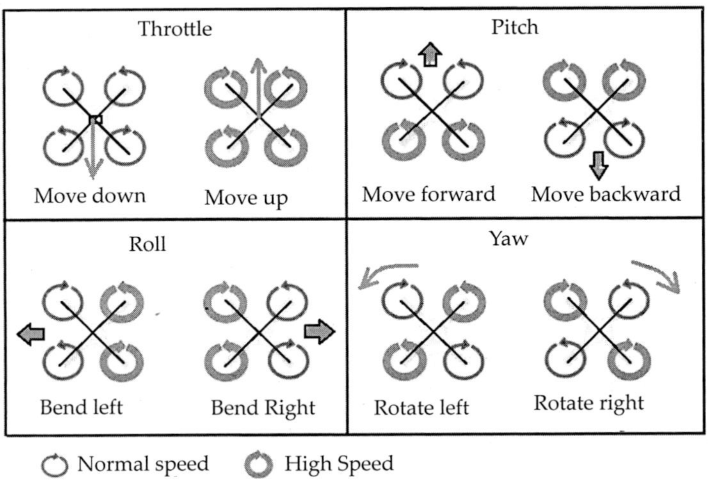

Figure: 2.15 Instructions for Operating a Drone and Managing a Quadcopter

FORCES AND MOMENTS ACTING ON A DRONE

Major Forces Acting on a Drone

When a drone ascends through the atmosphere, multiple forces exert influence upon it. The resultant force will determine its motion. Significant forces are exerted on a drone.

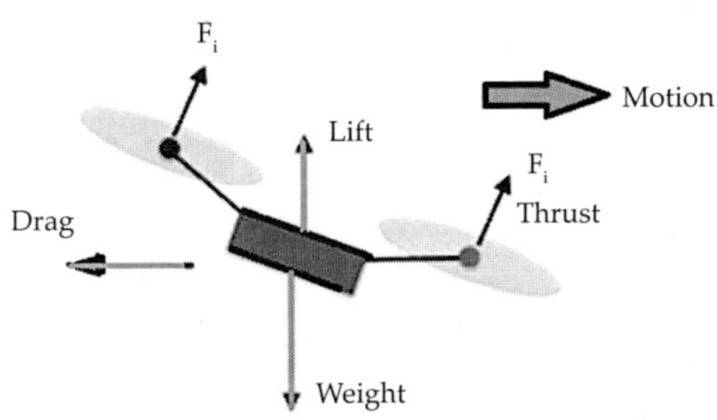

Figure: 2.16 Operational Principles of Drones and the Forces Exerted on Them

Weight

- The drone's bulk ensures that the gravitational force consistently works downward.
- The greater the drone's weight, the more power is need for its elevation and movement.
- Weight of the drone equals mass of the drone × gravitational acceleration

Lift

- The upward force exerted on the drone is referred to as lift.
- This force arises from pressure differentials across the drone in the vertical orientation. Thus, the velocity, dimensions, and configuration of the propeller blade determine the magnitude of the lift force.
- Elevation is crucial for raising the body against gravitational forces.
- All four propellers operate at high velocity to generate the necessary lift for the drone.

Thrust

- The force exerted on the drone in the direction of its movement is referred to as thrust. Nonetheless, for drone dynamics, it is oriented to the rotor plane.
- During hovering, the thrust is exclusively vertical. If propulsion is angled, the drone will pitch forward or backward.
- This force is crucial for propelling the drone in the intended direction at a consistent velocity.
- Two propellers have been equipped with high-speed capabilities to achieve the desired motion.

Drag

- The force opposing the drone's motion owing to air resistance is termed drag.
- This may result from pressure differentials and the viscosity of air.
- The aerodynamic configuration of the drone is chosen to minimize drag.

Rigid-Body Dynamics

- To compute the individual velocities and forces exerted on drones, it is essential to simulate the three-dimensional rigid-body dynamics.
- The initial phase involves identifying the reference coordinates, the direction of rotor speed, and the forces acting on the drones.
- For the rigid body, it is essential to account for the influences of aerodynamic, inertial, gravitational, and gyroscopic forces.
- **Aerodynamic Forces:** rotation of the propellors in air causes various forces such as friction and drag
- **Secondary Aerodynamic Effects:** blade flapping, ground effect, and local flow fields

- **Inertial counter torques:** gravitational forces acting at the center of drone affect the rotation of propellors
- **Gyroscopic effects:** change in the orientation of drone body and plane rotation of propellors.

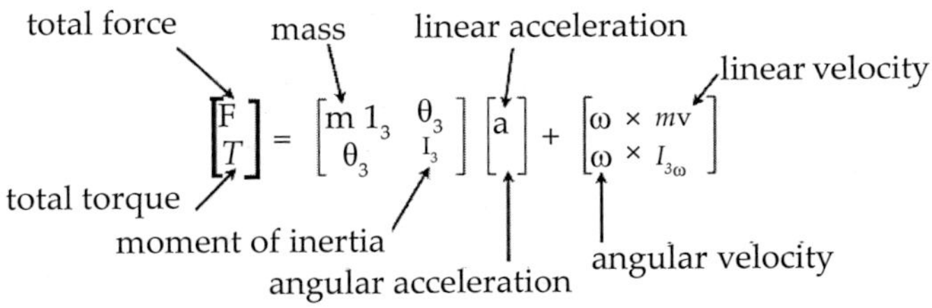

Figure: 2.17 **Based on Newton-Euler equations, all Forces and Moments Acting on a Quadcopter are Combined and Result in a Complete Model of the Drone Dynamics**

This physical model is useful to control the desired motion of the quadcopter

3

Various Sensors Used in Agricultural Drones

Population expansion, climate change, diminished rainfall, and escalating food demand are adversely affecting agriculture, resulting in alterations to crop practices. Consequently, to enhance yields and obtain precise data, it is imperative to implement modern precision farming techniques and deploy diverse sensor technologies.

- In light of the current circumstances and their adverse effects on conventional farming methods, agriculture must be conducted more wisely, employing advanced and cutting-edge technologies. This is the sole method to deliver a solution and address the incessant and expanding demands of the global population.
- Precision agriculture sensors are highly effective in agriculture as they relay data that enables farmers to monitor, enhance their goods, and stay informed about changes in the field and ecosystem.
- Advanced agricultural sensors efficiently identify animals, detect estrus, and monitor health, so enabling the isolation and treatment of ill cows through the identification, detection, and tracking of herds.
- Farmers can now utilize smart sensors in agriculture to monitor their crops remotely, assess their efficacy, manage pest infestations, and promptly implement measures to safeguard their crops from environmental threats.
- A sensor is a device that detects and reacts to specific inputs such as light, movement, pressure, temperature, or humidity, converting them into representations or signals that can be interpreted by people for further analysis and processing.
- They are frequently utilized in many applications, ranging from motion detection in security systems to temperature measurement in HVAC systems. They are also utilized in commonplace items such as smartphones, automobiles, and appliances.

- Sensors operate by identifying physical or chemical alterations in the environment and transforming them into electrical impulses. The sensor utilized is contingent upon the nature of the change being monitored.
- A temperature sensor identifies variations in temperature and transforms them into electrical signals that may be understood by the connected device.

Various sensor types are frequently employed in UAV remote-sensing applications within the agricultural and forestry domains. They can mostly be categorized as follows:

RGB CAMERAS

This involves use light to assess soil materials and monitor various light occurrences. These sensors can be affixed to autos, satellites, drones, or robots, facilitating soil reflection and the collection and analysis of plant color data.

Figure: 3.1 Drone RGB Camera

Optical sensors possess the capability to condition the clay, organic matter, and moisture characteristics of the soil.

RGB sensors are the predominant sensors utilized in UAV systems for precision agriculture. They can capture high-resolution photos and are typically more economical than other varieties. Besides being lightweight and user-friendly, they are also readily available. RGB photographs can be shot in various weather circumstances, including sunny and gloomy days, provided that specific camera parameters (shutter speed, ISO, etc.) are appropriately adjusted according to the weather to avoid underexposure or overexposure. The primary limitation of these sensors is their reduced spectral resolution, rendering them unsuitable for assessing vegetation-related indices that require spectral data from the non-visible spectrum. Consequently, RGB sensors are generally employed in conjunction with many other sensor types.

MULTISPECTRAL AND HYPERSPECTRAL CAMERAS

Spectral data is essential in evaluating many biological and physical attributes of crops. Multispectral or hyperspectral imaging sensors advantageously possess elevated spectral resolution criteria. UAV-captured images equipped with these sensors can yield data regarding the spectral absorption and reflection of plants across several bands. Diseased crops can be identified early because chlorophyll absorbs visible light in the red spectrum while reflecting near-infrared light significantly. The data in the NIR channel can thus be utilized to detect instances of sickness, even if they are not yet apparent in the red channel. Advancements in sensor manufacturing have led to the development of lightweight multispectral sensors that can be utilized on contemporary commercial lightweight UAVs. The MicaSense RedEdge and Parrot Sequoia are two sensors capable of capturing an extensive spectrum of wavelengths, including near-infrared (NIR). Utilizing such cameras enables the extraction of NDVI and several other aforementioned vegetation indices. Figure 2 illustrates the spectral range encompassed by each of the two cameras. Hyperspectral sensors have been enhanced to provide lightweight variants specifically designed for drones. These cameras encompass a broader spectrum range than multispectral sensors; nonetheless, they typically attain inferior spatial resolution. The HSC-2 is a sensor frequently employed for the classification and monitoring of vegetation. This lightweight hyperspectral camera produces up to one thousand spectral bands in the VIS-VNIR range, tunable from 400 to 1,000 nm.

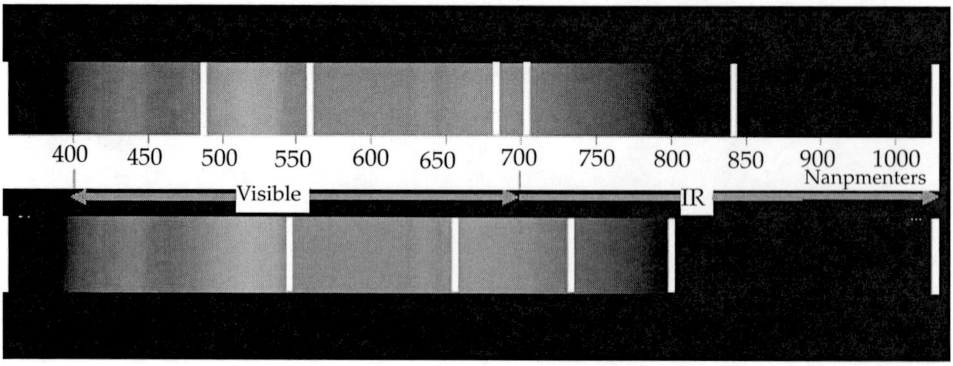

Figure: 3.2 Two instances of sensors capable of detecting a broader spectrum of wavelengths, including near-infrared (NIR).

THERMAL CAMERAS

All objects emit infrared light at specified wavelengths proportional to their temperature when they are above absolute zero. Thermal cameras focus on and detect radiation at specific wavelengths, which is then converted into a grayscale image to depict heat. Certain thermal sensors can analyze the received energy and convert it into the object's absolute temperature. In this instance, visuals may be rendered in color, with hotter items typically represented by red hues and colder ones by blue. Thermal UAV remote sensing has been effectively employed in agriculture for assessing plant health, predicated on the premise that healthy plants typically exhibits decreased dryness at reduced temperatures. Furthermore, these photos may undergo additional examination and processing to derive information beneficial for irrigation mapping, fertilizer and pesticide control, and yield estimation. Furthermore, thermal imaging can aid in evaluating a plant's maturity progression and can detect crop damage resulting from detrimental occurrences such as fires and similar disasters. FLIR is a prominent camera brand utilized in this area, providing the three primary wavelength ranges of short-wave, mid-wave, and long-wave infrared.

Figure: 3.3 Thermal Camera

LIDAR

The 3D point cloud is crucial for evaluating plant height, crop phenology, water movement, and additional factors. A more accurate estimation of these properties can be attained when a high-density point cloud is present. A UAV outfitted with Lidar is the optimal approach for obtaining a dense point cloud, owing to Lidar's exceptional precision and capacity for numerous returns. Biomass is conventionally assessed using quadrats that have not been defoliated through cutting or grazing throughout the growing season. Despite the data's accuracy, this conventional approach of AGB field sampling is time-intensive, expensive, and arduous. Furthermore, gathering AGB data poses challenges in extensive regions. Lidar data obtained from UAVs offers a distinctive chance to quantify three-dimensional vertical structures. Canopy height assessments depend on the return of laser pulses from vegetation and the ground, involving several reflections at various elevations. In recent years, digital terrain models (DTMs) and digital surface models (DSMs) have been generated from Lidar data to facilitate the monitoring of plant structures via UAVs and the development of canopy height models (CHMs). To ascertain the CHM, the DSM is evaluated at various intervals throughout the growth season, with the initial DTM calculated from the initial observations. Methods employing Lidar have emerged to determine the PAI based on canopy density. The primary advantage of these Lidar approaches is their sensitivity to both chlorophyllous and non-chlorophyllous plant elements.

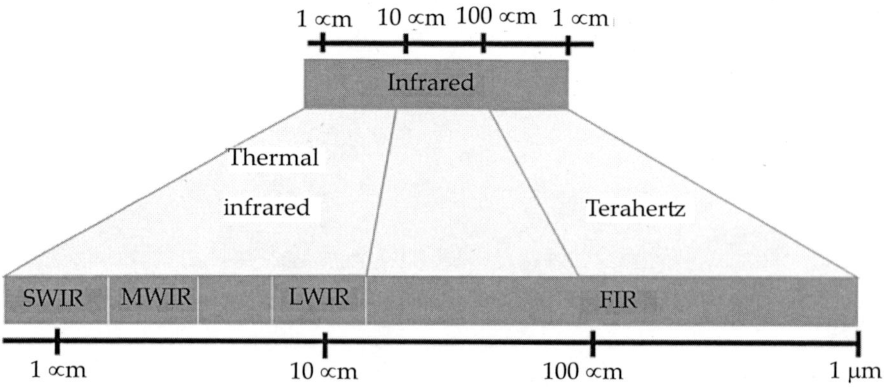

Figure: 3.4 The three primary wavelength ranges of FLIR cameras.

ADVANTAGES OF UTILIZING AGRICULTURE SENSORS

1. Precision agriculture with agricultural sensors facilitates the shift from conventional farming techniques to a data-centric methodology, enhancing resource distribution and maximizing yield.
2. Sensors facilitate water conservation and mitigate environmental effect by delivering real-time data for informed decision-making.
3. Advanced monitoring stations equipped with sensors collect climatic data, enabling you to track weather fluctuations and adjust your agricultural activities accordingly.

4

Applications of Different Sensors in Agriculture

In agriculture, sensors can monitor soil, water, vegetation, and environmental conditions. They assist farmers in enhancing crop quality, optimizing water usage, and making informed fertilizer choices.

CROP MONITORING

The primary function of remote sensing in agriculture is to assess crop health. Optical (VIR) sensing enables the observation of wavelengths beyond the visible spectrum, such as infrared; these wavelengths are highly responsive to crop vitality, damage, and stress. Recent advancements in this technology have enabled farmers to monitor their fields and execute prompt crop management decisions. Remote sensing for crop identification facilitates the detection of crops impacted by weather-related circumstances, pests, and other factors.

Pest Detection

Drones can administer insecticides and herbicides with greater precision, reducing chemical consumption and environmental repercussions. Drones can minimize chemical application by targeting specific regions impacted by pests or weeds, thereby safeguarding beneficial insects and mitigating the danger of pesticide resistance. Drones can also be utilized to deploy biological control agents, including beneficial insects or fungi, to address pest issues in an environmentally sustainable way.

Irrigation Management

A primary application of UAV technology in smart agriculture is crop irrigation management, which enhances water consumption efficiency by delivering the precise quantity of water at the most suitable location and time. UAVs equipped with suitable sensors can identify regions of a crop that necessitate supplementary irrigation. This positively impacts overall crop productivity. The thermal camera is among the most frequently utilized sensors for this application. Utilizing thermal imaging to assess the fluctuation of water potential in a field can enhance irrigation efficiency. Thermal photography from UAVs may be essential for detecting issues like faulty irrigation systems and inconsistent irrigation. The utilization of UAV thermal imaging enables the identification of heat stress prior to irreversible damage to the plant, as well as the detection of irrigation issues before symptoms become clearly apparent. In vineyards, UAV-based irrigation is employed in smart agriculture to optimize water distribution, facilitate consistent maturation, and thus improve product quality.

CROP INSURANCE

Global issues such as climate change and intensified precipitation are progressively leading to floods, which causes crops to die from insufficient oxygen to the roots and inadequate sunshine for the foliage. In these instances, farmers frequently file insurance claims for the incurred damage. Prior to the claim being honored, an insurance loss inspector must do a crop damage assessment. Historically, these tasks have been difficult to do and are often inaccurate due to their predominantly manual nature. UAVs provide the capability for agricultural damage evaluations since they facilitate image acquisition across extensive areas, hence minimizing time and costs, enhancing inspection precision, and aiding in effective documentation.

WEED MAPPING

Weed mapping constitutes a significant use of UAV-enabled precision agriculture. Weeds are undesirable flora that proliferate in agricultural fields and can cause various problems. They adversely affect plant growth by exacerbating competition for scarce resources such as water and space. The traditional approach to managing weeds involves applying pesticides to the crop field. Excessive pesticide application might result in herbicide-resistant weeds and adversely affect crop growth and productivity. Moreover, it poses a substantial threat to the environment via pollution, while also markedly elevating production costs. An alternative to this is site-specific weed management (SSWM), wherein herbicides are administered in a spatially varied manner instead of being uniformly sprayed throughout the entire area. This necessitates the development of a precise weed cover map that facilitates targeted herbicide delivery with UAV images. Task weed maps or zone maps illustrate soil types, geography, or pest distributions. Machine learning algorithms applied to UAV imagery enable the automatic identification of weeds and crops based on their shape, structure, and contrast. Herbicides are sprayed exclusively to the weeds, which benefits the crop and minimizes environmental damage.

CROP DAMAGE DETECTION

Unmanned Aerial Vehicles (UAVs) can generate three-dimensional digital maps of crops and evaluate several parameters, including crop height and Leaf Area Index (LAI). Moreover, UAV-derived crop image data can be analyzed using data processing methods to detect variations in plant biomass and health.

Consequently, diseases can be detected in their first phases, enabling farmers to implement measures to mitigate losses. UAVs primarily function by collecting data pertinent to crop health to identify potential infections in their early stages, prior to the emergence of visible symptoms. Furthermore, UAVs can be employed in infection management, enabling farmers to conduct targeted spraying and accurately monitor the intervention's progress.

Fertilizer on Crops

Nitrogen is the most widely used fertilizer globally; it is an essential element of chlorophyll, crucial for plant survival. Regulating the amount of fertilizer applied to crops advantages both the plants and the financial resources. Consequently, numerous farmers currently employ UAVs and advanced agricultural methods to gather data and map their fields, enabling them to supply essential fertilizers to their crops while minimizing nitrogen usage. The GAI is the predominant index utilized to determine the timing and location for nitrogen application. The resultant crop map is often color-coded based on GAI values, where lower GAI values signify a heightened requirement for nitrogen.

Flooded Crop Area Inspection

For insurance loss assessments, a UAV can be sent over the affected areas to collect data utilizing RGB or MS cameras. The data can be analyzed and presented with advanced technology to quantify crop loss. To assess the crop damage and generate a report for the farmer, the loss inspector should initially create a sequence of visualizations akin to an orthomosaic map. The loss inspector may also calculate the NDVI, which is efficient in identifying vegetation and bare soil, to assess the extent of damage. Upon creating the categorized NDVI map that illustrates the bare soil patches within the crop field, the inspector can utilize the map to pinpoint regions necessitating a more thorough evaluation of damage severity.

PAYLOAD INTEGRATION AND COMPATIBILITY CONSIDERATIONS

Payload integration denotes the meticulous installation of a payload (the principal scientific or operational apparatus) onto a launch vehicle or spacecraft, necessitating comprehensive evaluation of compatibility factors such as dimensions, mass, power specifications, electrical interfaces, thermal limits, and environmental conditions to guarantee mission success during launch and in space; critical compatibility aspects encompass structural integrity, electromagnetic compatibility, data communication protocols, and the payload's resilience to the launch environment.

The payload capacity of a drone, or its weight-carrying capability, should be taken into account while purchasing an agricultural drone. This is particularly applicable when the drone is employed for the application of fertilizers and pesticides on agricultural fields.

A drone with an increased payload capacity can transport additional equipment, like cameras for agricultural monitoring or tanks for the application of fertilizers and pesticides. A substantial payload capacity enables a drone to execute numerous missions without the need for frequent refueling.

The initial step to guarantee payload compatibility is to explicitly and precisely delineate the payload characteristics and requirements. This encompasses the payload mass, dimensions, configuration, center of gravity, inertia, structural integrity, thermal characteristics, electrical power, communication, and safety.

Factors that Affect Drone Payload Capacity

1. **Structural Design:** The dimensions and configuration of the frame influence the drone's payload capacity, as does the material of the frame. Fixed-wing drones has more robust construction to accommodate heavier payloads.
2. **Motor Arrangement:** Drones with additional motors, such as hexacopters and octocopters, can more effectively disperse the burden, hence enabling the lifting of bigger payloads. The positioning of the motor also guarantees enhanced lift and stability during flight.
3. **Propulsion:** Enterprise drones possess an increased payload capacity owing to the enhanced thrust generated by rotor dimensions and structural design.
4. **Power System:** Larger batteries provide enhanced lifting capacity and extended flying duration. Certain industrial drones have hybrid electric/gasoline powertrains, enhancing their lifting capacity.
5. **Remember:** A direct relationship exists between a drone's lifting capability and its maneuverability. Every drone possesses a designated center of gravity essential for stability and performance. Mounted payloads displace the center of gravity, modifying the drone's performance across the three axes: pitch, yaw, and roll. Native payloads typically address this issue, however third-party payloads may impair the drone's performance.

Large Payload Drones

Model	Payload Capacity	Max flight Time
Draganfly Heavy Lift Drone	67 lbs / 30.kg	55 minutes
Hercules 20	33 lbs / 15 kg	40 minutes
Perimeter 8 UAV	22 lbs / 10 kg	5+ hours, thanks to an internal combustion two stroke engine
IF1200	19.1lbs / 8.6 kg	43 minutes

Is it Possible to Increase Drone Payload Capacity?

Indeed, technically affirmative. Enhancing the drone's power and thrust will augment its payload capacity. In actuality, this would necessitate comprehensive customisation of the airframe, rotor, and propeller. Further performance testing is required. It may be advantageous to select a high payload drone from the outset.

Popular Types of Drone Payloads

The drone payload market is saturated with excellent native and third-party alternatives.

- Native payloads are specifically designed by the drone manufacturer to seamlessly integrate with particular drone models. Some are intrinsically integrated (e.g., camera systems). Some are mountable and interchangeable, such as various ultrasonic probes.
- Tools and accessories compatible with drones, created by entities other than the drone manufacturer, are referred to as third-party payloads. They can be elevated to a payload drone but may require customization and integration for adequate controls.
- Presented below are the premier drone payloads across the following categories:
    - RGB cameras
    - Infrared cameras
    - Ultrasonic transducers
    - Multispectral detectors
    - Combustible gas detectors
    - Density Functional Theory gauges
    - LiDAR technologies
    - Magnetometers
    - Lightning protection system evaluators
    - Devices for detecting radiation

RGB Cameras

Premier drones are outfitted with high-quality color cameras. Models may be equipped with HD visual inspection cameras, multifocal photogrammetry cameras, and wide-angle lenses for land surveying, contingent upon the specific application.

Recent drone models are equipped with cameras utilizing Complementary Metal-Oxide Semiconductor (CMOS) sensors. Due to on-chip processing and individual pixel readouts, CMOS sensors exhibit reduced power consumption, hence prolonging drone flying duration. They provide accelerated image capture and readout velocities, an expanded dynamic range, enhanced clarity, and an elevated frame rate.

Furthermore, CMOS facilitates the seamless incorporation of additional circuitry on the sensor chip, like amplifiers, noise reduction mechanisms, or Bayesian filters to enhance film quality. The ANAFI Ai drone incorporates a 48 MP 1/2" CMOS sensor, featuring a Quad Bayer color filter pattern, to achieve enhanced image resolution and an expanded dynamic range.

Figure: 4.1 ANAFI AI Lens has six Aspherical Elements, Designed to Minimize Optical Flare. Origin: Parrot

The zoom range is a significant attribute of drone cameras. When utilizing drones for asset integrity assessment, seek models equipped with a hybrid zoom camera that integrates both optical and digital zoom capabilities for enhanced range. A zoom range of 10X to 20X is quite conventional.

Top inspection drones, such as the DJI Mavic 3 Enterprise, provide a 56X zoom capability. To achieve a clearer image, use drone cameras featuring a bigger sensor size (e.g., 1-inch, APS-C), as they provide superior low-light performance and dynamic range. Larger apertures (e.g., f/2.8 or lower) and elevated ISO sensitivity enhance footage acquisition in low-light environments.

In summary, numerous native and third-party camera systems for drones are available, making the selection dependent on the required specifications.

Infrared and Thermal Cameras

Electro-optical (EO) and infrared (IR) camera payloads are widely utilized accessories. Thermal imaging is advantageous in search and rescue operations and drone surveillance missions for the rapid identification of human presence. Thermal drones assist in firefighting and environmental surveillance.

- In asset management, EO/IR systems facilitate the detection of temperature fluctuations, which signify damage, wear, or performance irregularities. Thermal drone inspections of power lines facilitate the identification of overheated lines and components, as well as the detection of anomalous heat patterns in transformers, all of which signify potential failure.
- Irregularities in thermal signatures may also signify obstructions in heat exchangers and industrial boilers. In the oil and gas industry, thermal drones facilitate pipeline inspections for material leakage. The Skydio X10 is equipped with an integrated FLIR Boson sensor and CMOS imaging technology, delivering radiometric measurements with a sensitivity of about 30 mK.
- The DJI Zenmuse H20T is a favored payload option, integrating a 20 MP zoom camera, a 12 MP wide camera, a 1200 m laser rangefinder, and a 640×512 px radiometric thermal camera. Additionally, exceptional thermal drone cameras are available from Gremsy. The updated VIO system integrates a 640×512

radiometric thermal camera, a 4K zoom sensor, and a built-in 2400m laser rangefinder.

Ultrasonic Teransducer

Traditionally, NDT was accompanied by logistical and staging constraints that hindered convenient physical access to the test structure. The preparatory costs surpassed those of conducting the inspection.

Specialized non-destructive testing (NDT) drones, such as the Voliro T, obviate the necessity for scaffolding or lifting apparatus in asset evaluation. The Voliro tricopter, designed for six-axis movement, can securely hover near structures while exerting 30 N of stable force and several Nm of torque to conduct point-to-point tests using drone-adapted ultrasonic testing probes.

- Ultrasonic transducer (UT) employing Echo-to-Echo and Pulsed Echo techniques for wall thickness measurements, achieving a resolution of 0.02 microseconds (0.06 mm at 5900 m/s). It includes a practical couplant applicator.
- The high-temperature UT payload possesses same qualities but has been fortified to operate within a temperature range of up to 260 °C/500 °F, enabling it to gather data on operational assets, such as identifying corrosion on a functioning flare stack.
- The Electromagnetic Acoustic Transducer (EMAT) thickness gauge offers a measurement range of 2-150 mm on corroded and contaminated surfaces, where the application of couplant is problematic.
- Utilizing Voliro T NDT payloads, inspection teams reduce asset inspection time by 2X-4X while also gathering double the amount of data. Mistras Group conducted an investigation of a 180 ft flare stack in under two hours, acquiring more than 200 thickness measurements. In a separate instance, Bilfingers reduced its client's operational expenses by almost $165,000 (including labor, scaffolding, and equipment) with Voliro, accomplishing a stack inspection in under 1.5 days rather than the customary 25 days.

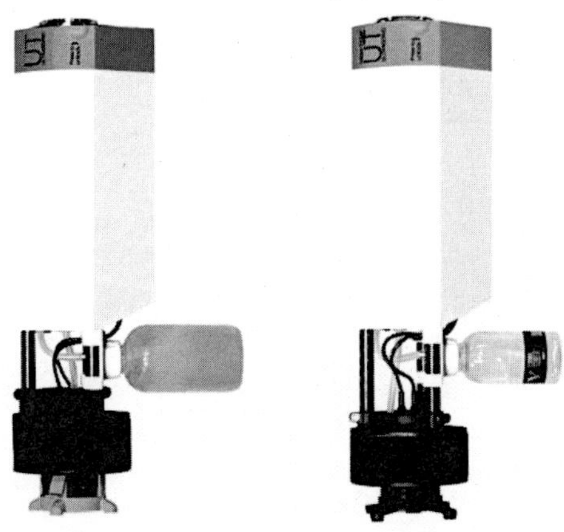

Figure: 4.2 The conventional UT probe and the High-Temperature UT probe. Obtain the specifications.

Multispectral Sensors

Multispectral sensors are highly beneficial in precision agriculture and environmental monitoring. Drones traverse larger areas more rapidly than foot or vehicular patrols. Plus is less expensive than chartered aircraft. Multispectral sensor payloads enable the frequent gathering of vegetation index data to identify regions under stress from pests, diseases, or water deficiency.

Similarly, multispectral cameras facilitate the acquisition of spectral signatures of prospective mineral resources and the mapping of topographical characteristics. Multispectral drones assist in evaluating environmental consequences and monitoring stockpile management efficiency at active locations to prevent accidents.

When purchasing multispectral cameras for drones, evaluate the quantity of supported spectral bands, wavelengths, and bandwidths. Models exhibiting elevated spectral and spatial resolution yield clearer images with reduced distortion. An adaptable sensor range facilitates superior image capture in low-light conditions. Ultimately, a global shutter eliminates motion blur in video recordings.

The premier multispectral sensors for drones include:

- **Senetra 6X Multispectral:** The lightweight sensor, weighing only 250 grams (8.8 ounces), integrates five 3.2MP global shutter multispectral imagers and one 20MP RGB imager into a single unit.
- **RedEdge-P:** A system complying with NDAA, with five narrow multispectral bands equipped with scientific-grade filters and output resolutions of 2 cm / 0.8 in at a distance of 60 m / 200 ft.
- **MicaSense Altum:** A combination of a 12 MP high-resolution panchromatic sensor with a Boson 320 x 256 heat sensor with a pixel size of 33.5 cm.

Gas Detectors

The optimal drones for the oil and gas industry are equipped with integrated or attachable gas detectors. Inspection drones can be readily equipped with electrochemical sensors for the detection of carbon monoxide, nitrogen dioxide, and hydrogen sulfide, or with infrared sensors for measuring infrared light absorption.

However, for more accurate measurements in sustainability reporting, a laser-based gas detector may be preferable. The most recent versions employ absorption spectroscopy to detect and quantify gas content. Tunable Diode Laser Absorption Spectroscopy (TDLAS) can identify extremely low gas concentrations with excellent accuracy and at greater distances than alternative approaches.

The gas detection payloads we propose:

- **FLIR MUVE C360:** Mountable multi-gas detector (CO, Cl_2, O_2, NO_2, H_2S, SO_2, LEL) including an incorporated snorkel that extends beyond the propellers to sample undisturbed air. Operated remotely using a USB-A accessory (tablet) and UAS power port with serial connection (C360).

- **Laser Falcon Methane Sensor:** A Class 2 laser for the detection of methane and methane-containing gases at distances of up to 60 meters or 100 feet. Provides automatic GPS time synchronization for report production, including time, leak concentration, GPS locations, and other data.
- **U10 Methane Sensor:** A methane detector utilizing TDLAS identifies gas presence at distances up to 100 m (329 ft) with concentrations below 5 ppm.

Dry Film Thickness Gauges

Routine examinations of protective coatings are essential for mitigating corrosion on metal structures, preventing leaks in storage tanks, and averting overall degradation. Correct application and prompt re-coating are mandatory compliance requirements across various sectors.

- Traditionally, digital DFT gauges are portable devices (such as the reliable Elcometer equipment), necessitating inspectors to ascend for measurements. Voliro's solution enables teams to remain on the ground while acquiring high-fidelity spot measurements.
- Our DFT probe, adhering to relevant EN ISO and ASTM B standards, offers dual measurement modes of 0-1500 μm for magnetic induction and 0-700 μm for eddy testing. The measurement ranges are 0 to 1.5 mm / 0 to 60 mils and 0 to 0.7 mm / 0 to 30 mils, respectively.
- Voliro's omnidirectional mobility and enhanced situational awareness provide a secure method for checking coatings prior to drying. EnerTrans evaluated over 10 transmission towers in a single day with a reduced crew, immediately following the coating application, thereby conserving time on subsequent visits and equipment commissioning.

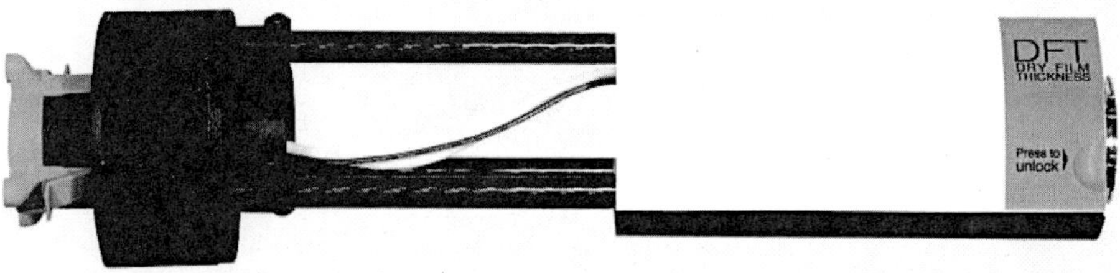

Figure: 4.3 DFT Probe for Inspection Drones

LIDAR Systems

Light Detection and Ranging (LiDAR) devices enhance location precision for asset photogrammetry and field surveys utilizing drones. Conventional photogrammetry utilizes an RGB camera to obtain a sequential series of images for 2D/3D modeling. This method may overlook subtle elements such as electricity wires or topographical characteristics obscured by foliage.

LiDAR systems employ laser pulses to delineate assets. The distance data, along with the sensor's position and orientation provided by GPS and IMU, establishes the reflection point of each laser pulse. Collectively, individual points form a high-density point cloud, representing a model of three-dimensional space.

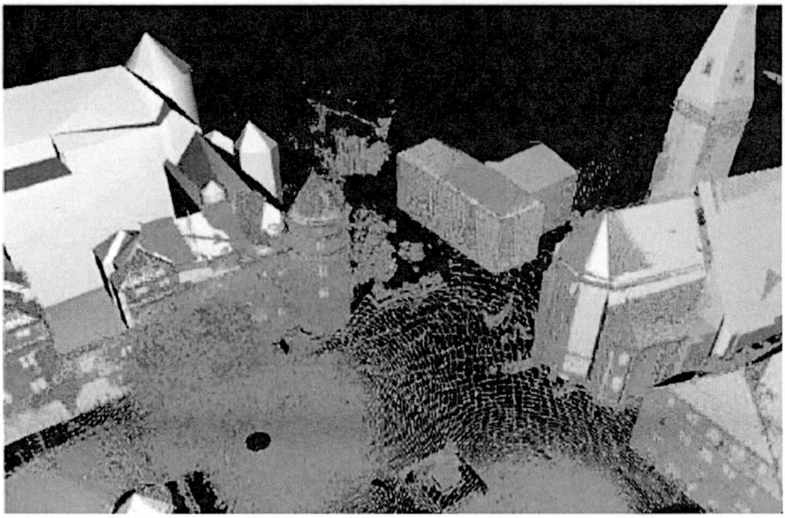

Figure: 4.4 Sample 3D Point Cloud of a City

In contrast to photogrammetry, which depends on picture matching and triangulation to deduce distances, LiDAR offers precise distance measurements, resulting in enhanced model accuracy. There is a reduced reliance on lighting conditions. LiDAR mapping can be conducted at night, in high-contrast illumination, or on homogeneous surfaces devoid of distinctive characteristics, such as monochromatic sand dunes.

Nonetheless, LiDAR models do not possess the photorealism characteristic of photogrammetry models. Consequently, numerous drone manufacturers integrate RGB and LiDAR into a singular sensing system. The DJI Zenmuse L2 payload has a Livox LIDAR sensor, a high-precision IMU, and a 20MP 4/3 CMOS RGB mapping camera. The system can identify objects from 250 m (820 ft) at 10% reflectivity and 100 klx, and up to 450 m (1480 ft) at 50% reflectivity and 0 klx.

YellowScan offers an impressive array of LiDAR systems compatible with drones. Voyager is a favored option, functioning at depths of up to 440 meters (1444 feet) with a precision of 0.5 centimeters (0.2 inches) and a centimeter-level accuracy range. It is also compatible with several RGB cameras. Surveyor Ultra is a lightweight, cost-effective alternative, providing 3 cm / 1.2 inch precision and 2.5 cm / 1 inch accuracy at elevations up to 140 m / 460 ft.

Magnetometers

Drone-mounted magnetometers enhance the depth of ground surveying. Magnetic field measurements enable the detection of subterranean pipelines and utilities during building activities.

Alternatively, pinpoint prospective exploration locations for extraction. MWH Geo-studies utilized UAVs equipped with magnetic payloads globally, spanning regions from the Andes to the Arabian Peninsula, to conduct mineral exploration studies. The company produced comprehensive orthophoto and DSM maps including more than 8,000 square kilometers.

The team favors laser-pumped Cesium vapor magnetometers for drones due to their superior sensitivity, exceptional precision, and capability for real-time data gathering in dynamic settings. They do not necessitate a physical surface with a defined area, allowing the drone to hover at a greater altitude to prevent collisions.

Quantum magnetometers for drones are an innovative device now in development by CanadianSBQuantum. The team is developing a drone-compatible magnetic gadget, energized by quantum impurities and diamonds. The sensor's core is a laboratory-cultivated diamond, incorporating carbon and nitrogen to induce spin, thereby rendering it a minuscule magnet. The amplitude and orientation of its magnetic field measurements yield exceptional precision without any null points.

Nonetheless, there is a plethora of outstanding classical magnetometers for drones, such as:

- **MagArrow:** A self-sufficient magnetometer system, incorporating a lightweight protective casing, MFAM magnetic sensors, GPS, and an inertial measurement unit (IMU). The MFAM, a dual-sensor module, exhibits a remarkable sensitivity range of 0.005 nT/√Hzrms and operates within a range of 20,000 nT to 100,000 nT.
- **MagNIMBUS:** QuSpin QTFM Gen-2 atomic total-field magnetometer sensors, mounted on a mechanically self-retracting arm for operation in proximity to the ground.
- **MagDrone R3:** A kit comprising a sensor tube equipped with two integrated 3-axis Fluxgates (each with a resolution of 150pT), a data logger with an internal SD card, rechargeable batteries, and a built-in GPS. Demands a drone payload capacity of no less than 1 kg/2.2 lbs.

Lighting Protection System (LPS) Testing Tools

Lightning protection systems (LPS) avert billions in possible damages each year, provided they operate effectively. Traditionally, LPS testing necessitates rope access or the construction of lifting platforms, resulting in extended downtime, particularly for wind turbines. A drone equipped with a payload can resolve those concerns.

The Voliro T, equipped with an LPS testing probe, diminishes the examination duration to an average of 20-30 minutes. The rotor of wind turbines can be halted at any position. Voliro's angular EN 62305-3 LPS probe is capable of interfacing with oxidized lightning receptors for elevated readings. The measured current is 0.30 A (for resistances less than 20 Ω).

Utilizing Voliro, Skyspecs, a renewable energy asset management firm, conducts inspections of 10-12 wind turbines daily with less personnel. Our high-precision LPS probe identifies three to four times more problems, resulting in enhanced wind turbine maintenance.

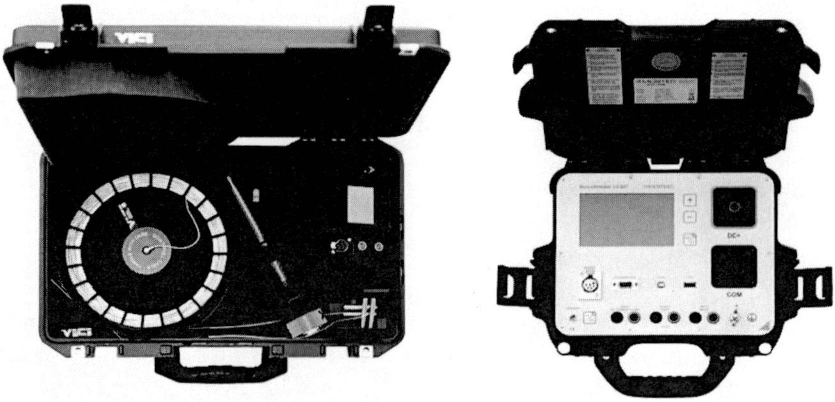

Figure: 4.5 Voliro LPS Testing Apparatus for Unmanned Aerial Vehicles

Radiation Sensors

Operational safety is essential in industries managing naturally occurring radioactive materials (NORM) and their derivatives. Radiation sensors fitted to drones provide remote measurement and monitoring of dosage fields at nuclear, mining, chemical, and waste management sites. Utilizing drones to identify and chart radiation hot zones enables teams to enhance intervention strategies and reduce human exposure.

Drones are also beneficial for conducting decommissioning surveys at defunct nuclear facilities or locations containing radioactive waste. Sellafield, a decommissioned UK nuclear facility, employs Elios drones equipped with a radiation detecting payload to inspect and create 3D maps of all aboveground and subsurface structures, including 1,000 buildings, prior to permitting personnel access.

Additional superior radiation sensors for drones include:

- **Kromek SIGMA 25/50:** A lightweight device featuring a gamma detection capacity of 32.8 cm^3 / 2 in^3 and an output of 54 photons per keV. TN15 is an exemplary model for neutron detection that operates at room temperature without the need for cooling.
- **SPIR-Explorer Sensor:** Integrates two scintillator detectors for radiation, providing a dose measuring range of 0.001 μSv/h to 10 Sv/h (0.1 μR/h to 1000 R/h). Additionally provides real-time nuclide detection and count rate for each nuclide.

Other Types of Drone Payloads

In addition to the aforementioned, drones can also be equipped with various payloads such as:

- Delivery systems such as baskets, droppers, and boxes equipped with drop-off devices for commercial delivery or the distribution of medical supplies.

- Agricultural payloads, comprising crop sprayers, fertilizer applicators, seed dispensers, and crop health monitoring sensors (e.g., hyperspectral or near-infrared sensors).
- Communication devices such as radio relays and mobile network extenders, affixed to tethered drones to enhance connection in remote regions.
- Public safety instruments such as floodlights, loudspeakers, emergency beacons, smoke detectors, or explosives detection sensors, intended to identify hazardous substances.

Things to Consider When Choosing a Drone Payload

To identify the optimal equipment available, consider:

- **Principal Application:** Identify essential hardware specifications for your application (e.g., high-definition video capture, temperature sensitivity, ultrasonic assessment). Analyze payload attributes including resolution, precision, detection range, and others.
- **Compatibility:** Verify that the payload is compatible with the drone's mounting system and control interfaces.
- **Mass:** Avoid exceeding the manufacturer-specified payload weight limit to prevent performance complications.
- **Power Specifications:** Confirm that the drone can provide sufficient power to the payload without jeopardizing battery depletion.
- **Resilience:** Select durable payloads that are resistant to temperature and weather conditions when the application necessitates such specifications.
- **Management of Data:** Verify that the payload is compatible with data streaming and/or possesses adequate onboard storage capacity. Verify the compatibility of your analytics software with the payload data format(s).
- **Instruction:** Advanced payload systems may necessitate additional pilot training. Select alternatives that provide ample technical documentation and training support.

Use of UAVs in Smart Agriculture

Currently, UAVs outfitted with imaging sensors are significantly contributing to digital smart agriculture. This technique can be a pivotal factor enhancing agricultural accuracy, augmenting crop productivity, and emphasizing targeted crop protection.

Unmanned Aerial Vehicles (UAVs) can be outfitted with various data collection sensors, including RGB, multispectral, hyperspectral, thermal cameras, or Lidar. Plants exhibit variable levels of light reflection based on their chlorophyll concentration and biomass. A UAV equipped with a multi- or hyperspectral camera quantifies the solar energy reflected across several spectral regions. The energy reflected from plants and the relationship among various spectral bands might facilitate the indexing of crop health status. Numerous vegetation indices exist, including the prominent Normalized Difference Vegetation Index (NDVI), which utilizes the near-infrared (NIR) band. The NDVI is typically employed to signify leaf covering and plant vitality, with elevated NDVI values reflecting a substantial biomass concentration. Consequently, NDVI and other vegetation index maps has significant potential for enhancing agricultural food production sustainably.

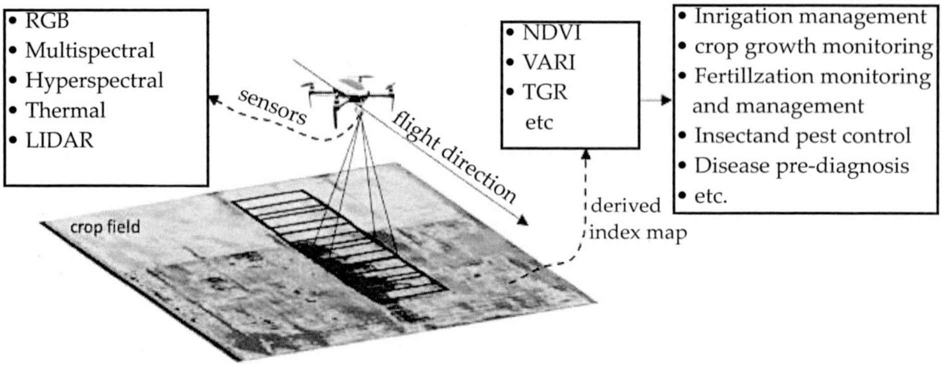

Figure: 4.6 A Depiction of UAV Utilization in Intelligent Agriculture.

Thus, spectral data obtained from a UAV can be analyzed to determine if the photographed crop is fresh or desiccated and to identify regions exhibiting initial indications of pest or disease infection. Furthermore, this spectral data, when integrated with three-dimensional information from sources like digital surface models (DSMs) and hydrological maps, can enhance the effectiveness of processes such as irrigation or pesticide application while minimizing the necessary inputs. Figure 1 delineates the application of UAVs in smart agriculture and illustrates the relationship among sensors, vegetation indices, and the corresponding applications as detailed below.

Vegetation Indexes and Measurements

Numerous spectral vegetation indices have been proposed and delineated to assess plant characteristics. Invisible specifications can be transformed into visible indices and metrics through the creation of suitable color maps or indicators. Instances of such indices comprise:

- **Leaf Area Index (LAI):** This refers to the projection of the one-sided green leaf area per unit of ground area. The Leaf Area Index (LAI) is a crucial indicator of crop growth condition. Estimation can be performed by ground-based techniques or remote sensing. Current optical remote sensing methods for estimating LAI are categorized into three types: empirical associations with vegetation indices such as NDVI, inversion of canopy reflectance models, or a combination of both approaches.
- **Normalized Difference Vegetation Index (NDVI):** This metric is extensively utilized in precision agriculture to evaluate the spatial variability of the anticipated Leaf Area Index (LAI). The NDVI is typically employed to signify leaf coverage and plant vitality, with elevated NDVI values indicating a greater biomass concentration.

- **Triangular Greenness Index (TGI):** This index utilizes RGB photographs, depending on reflectance values in the visible spectrum to signify chlorophyll concentration in regions with dense leaf coverage.
- **Visible Atmospherically Resistant Index (VARI):** This index is obtained from RGB photographs to assess leaf covering. The VARI index is employed to assess vegetation components that exhibit minimal sensitivity to atmospheric conditions.

  Above Ground Biomass (AGB) encompasses all biomass present in living plants above the soil, including stems, branches, seeds, and foliage. As above-ground biomass is predominantly observable in carbon pools (such as seas), AGB serves as a significant indicator of alterations or advantages associated with carbon mitigation. The AGB serves as a crucial metric for tracking greenhouse gas emissions and associated initiatives.
- **Fractional Vegetation Cover (FVC):** This index encompasses the ground layer (e.g., grasses), the mid-layer (e.g., mid-sized trees), and the canopy layer. The NDVI values of bare soil reflectance (NDVIs) and full vegetation reflectance (NDVIv) can be utilized to ascertain the Fractional Vegetation Cover (FVC).
- **Green Area Index (GAI):** This metric encompasses all green components of a plant, including leaves, stems, and reproductive structures. The complimentary Brown Area Index (BAI) denotes non-verdant regions within the canopy. The Plant Area Index (PAI) is a generic metric that include both green and non-green components.

5

Principles of Flight Planning for Agricultural Drone

INTRODUCTION TO FLIGHT PLANNING FOR AGRICULTURAL DRONE

Flight planning for agricultural drone operations entails selecting the appropriate drone, preparing the drone and its equipment, and devising a flight trajectory. Drone flight planning software enhances safety and ensures compliance.

Why is Drone Flight Planning Important?

Effective drone flight planning establishes the groundwork for effective mapping and surveying operations. It entails formulating a comprehensive flight plan for the drone, guaranteeing that data acquisition is precise, effective, and secure. Inadequate preparation exposes operators to the risks of missing data, overlapping imagery, or delays in mission execution. Principal advantages comprise:

- **Precise Data Acquisition:** Strategically organized flights provide uniform overlap among photos, essential for producing high-quality orthomosaics and 3D mappings with a drone.
- **Optimized Resource Utilization:** Efficient flying trajectories reduce energy consumption and time allocated for data acquisition.
- **Improved Safety:** Pre-arranged missions circumvent impediments and comply with airspace laws, thereby mitigating dangers.

COMPONENTS OF DRONE FLIGHT PLANNING

Defining the Mission Objective

Each mapping or surveying endeavor commences with a clear comprehension of the precise purpose. The mission aim dictates the necessary data and the appropriate tools for land surveys, construction site inspections, or agricultural mapping.

Choosing the Right Equipment

Choosing the appropriate drone, sensors, and flight controller software is essential. The combination must correspond with the mission's requirements, whether it necessitates RGB cameras for photogrammetry, multispectral sensors for agriculture, or LiDAR for high-precision surveys.

Determining Flight Parameters

Essential elements for efficient drone flight planning encompass height, overlap, and speed. Altitude influences image resolution and area coverage; lower altitudes yield greater detail but necessitate increased flight duration. Overlap guarantees precise image stitching and analysis. Velocity influences image sharpness, with reduced speeds advised for obtaining intricate information.

Compliance and Safety

Compliance with local aviation regulations, airspace authorizations, and restricted flight areas is obligatory. Safety protocols, including fail-safe and return-to-home (RTH) configurations, must be established.

SELECTION OF APPROPRIATE FLIGHT PARAMETERS

These characteristics are pertinent to all regular flights for orchard mapping and to the majority of thermal flights for orchard mapping. A distinct part delineates the precise settings to employ during the thermal flights for orchard mapping (i.e., operating with the Altum).

Parameters

Table : 5.1 Flight Parameters Comparison for DJI Phantom 4, Matrice & Similar Drones vs. DJI M3M/M3T

Parameter Category	DJI Phantom 4, Matrice & similar sized drones	DJI M3M/M3T	Notes
Overlap	70%		
Sidelap	70-75%		
Weather	Full sun or full cloud cover		Thermal data will be compromised if flights are conducted in scattered clouds
Speed	12m/s or 40 ft/s	11m/s or 36 ft/s	At a lower flight height the flight speed should be reduced to avoid coverage issues
Height	Small/Medium trees (0.5m - 3m): 100m or (1.5ft - 10ft): 330ft		Ideally only fly lower for smaller trees. Flying lower will decrease flight efficiency and increase the amount of data collected, uploaded, and processed
	Large trees (3m & above): 120m or (10ft & above): 400ft		
Camera settings	Use mechanical shutter when applicable	DJI M3M: Dewarp Disabled	

Flying Tips

- Consistently ascend from the apex.
  - Ensure you possess a larger vehicle capable of navigating the challenging terrain of farms and facilitating easy transit between blocks.
  - Ensure that the elevation change per flight should not exceed 30 meters.
- Travel between 9:30 AM and 3:30 PM. Smaller trees or vines may be transported starting at 9:30 AM. Larger trees will be transported nearer to solar noon. Commence and conclude the day with diminutive trees to mitigate the likelihood of shadows. Larger trees will cast more extensive shadows for an extended duration during the day.
- Avoid flying in the presence of rolling or dispersed clouds. Flight is permissible under complete cloud cover.
  - Ensure the DJI drone camera is configured to default settings.
  - Ensure that your MicaSense camera is equipped with the most recent firmware update. Verify the most recent update here.
- During layovers, transfer your data to your computer and verify the absence of damaged or missing photographs. It is advisable to verify that all items are accounted for prior to departing from the farm. It is undesirable to get home and attempt to upload just to discover missing photographs or inadequate coverage.
- When choosing a boundary for flight, ensure to include a buffer to guarantee enough coverage. Either extract the blue dots or navigate with an expanded flying path. The MicaSense camera frequently omits a photograph throughout its rotation.
- Instances of the power plug on the MicaSense camera experiencing disconnection have occurred. We advise affixing the cord with putty or hot glue to prevent vibrations during flight from causing the camera to power off and miss capturing photographs.
  - To ensure sufficient MicaSense MSP coverage from your flights, upload all photographs to the Aeroview uploader in the field.
- Should the boundary be triangular, please redraw it as seen below.

Drone Altitude - The GSD Variable You Control

Because most of the image quality variables are predetermined by your drone setup, the main variable you can (or should) control is altitude. But changing altitude has important workflow implications.

The Time vs Detail Trade-Off

The primary variable you can adjust regarding image quality, given that other factors are predetermined by your drone configuration, is altitude. However, altering altitude has significant ramifications for workflow.

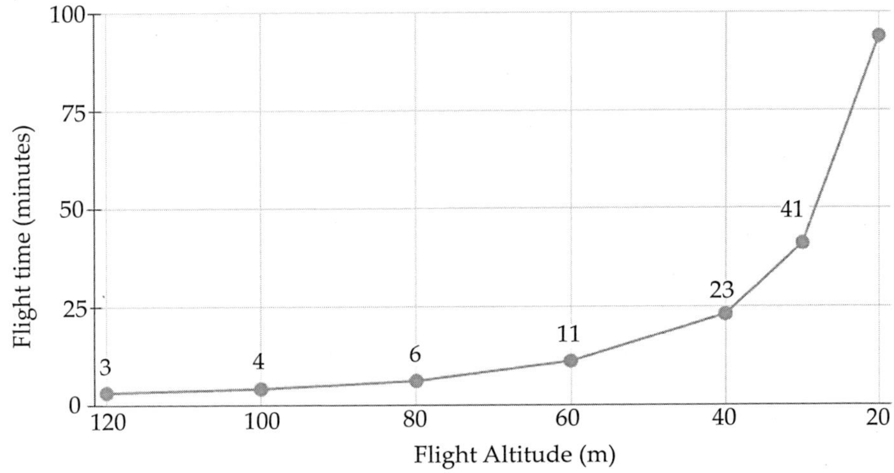

Figure: 5.1 The Duration of the Flight Significantly Escalates at Reduced Altitudes. Covering a 10 ha Field at 80-120 m Mltitude Requires Only a few Minutes, However Flying at 20 m Altitude May Take up to 1.5 Hours.

Extended flight duration results in an increased number of photographs, necessitating greater storage capacity and prolonged processing time.

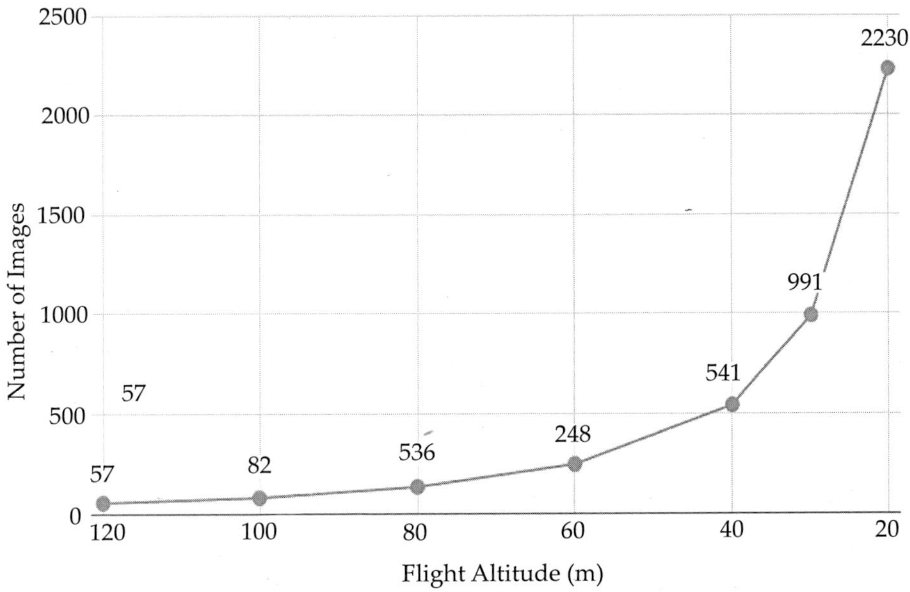

Figure: 5.2 The Quantity of Collected Photographs About Doubles with Every 20 Meters of Decreased Altitude, Akin to the Flight Duration. Increased Pictures Necessitate Greater Storage Capacity and Extended Processing Duration.

We advise selecting the maximum altitude (minimum flight duration/fewest images) that still provides the necessary detail for your objective (GSD).

Checking Camera Specs

Numerous contemporary high-resolution cameras achieve a low ground sample distance (detail) at unexpectedly elevated elevations. Therefore, familiarize yourself with the specifications of your drone camera. For a ground sample distance (GSD) of 0.50 cm/px, a typical requirement for plant counts, it is necessary to operate at an altitude of around 20 meters (65 feet) while utilizing a DJI Phantom 4 Pro drone. Utilizing the more sophisticated and costly DJI Zenmuse P1 camera on a Matrice 300 drone allows for the attainment of equivalent ground sampling distance (GSD) at an altitude of 40 meters (130 feet), hence significantly conserving flight time.

Online GSD calculators, such as those offered by Propeller Aero, may readily verify the Ground Sample Distance based on your drone configuration and intended altitude. It is prudent to verify that your flight plan will yield the necessary data.

Verify Drone Flight Speed

A complex effect of altitude is the drone's ground velocity. Extensive photos captured at low altitudes necessitate reduced flight velocities to prevent motion blur. Flight software, such as DJI Ground Station Pro, autonomously adjusts the drone's velocity to ensure photographs are clear and precise.

Agricultural Drone Altitude Best Practices

What is the optimal altitude for operating an agricultural drone?It is fundamentally an issue of 'What GSD is required?'Once the Ground Sampling Distance (GSD) is established, the parameters of your drone camera dictate the altitude for your operation.

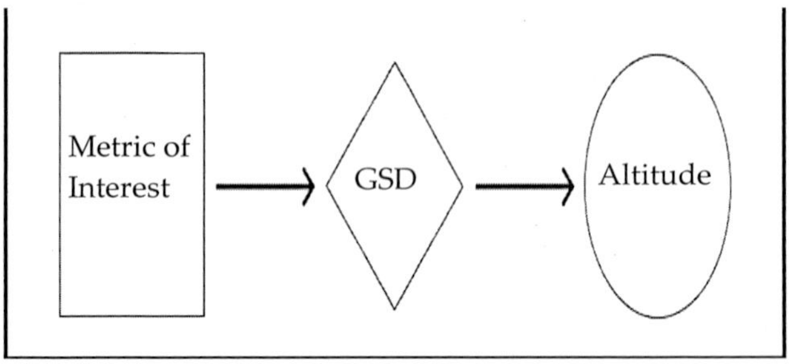

Figure: 5.3 The Decision-Making Flow

After processing thousands of flights at Solvi, we've seen a lot of trial and error. We've found that most drone missions fall into one of two categories: low or high detail (GSD).

Drone Altitude for Field Mapping or Precision Ag Management (Low Detail)

Establishing field boundaries, delineating field elevation, or extracting experimental plots does not necessitate excessive precision. Likewise, if your drone operations pertain to crop planning, diagnostics, crop monitoring, or prescriptive applications, a comprehensive perspective is adequate.

Under optimal conditions, the majority of drone cameras, including multispectral variants, can effortlessly achieve a Ground Sample Distance (GSD) of 3-10cm at elevations of 80-120m.

- **Target Ground Sampling Distance:** 3-10 cm
- **Elevation:** 80-120 meters (250-400 feet)
- Prevalent applications:
    - Boundary demarcations
    - Topographical cartography
    - Infrared thermography
    - Extract plots with well delineated borders
    - Vegetation indices for plant health assessment
- Weed identification for targeted herbicide application
    - Variable rate prescription application maps
    - Quantification of vegetation on substantial entities such as trees

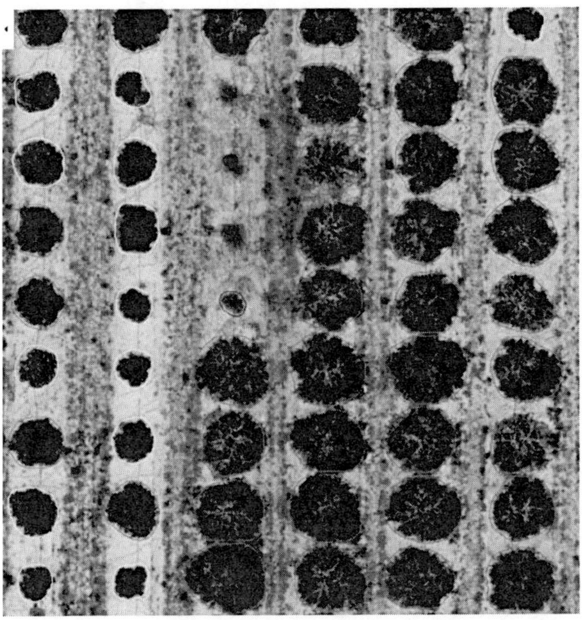

Figure: 5.4 Applications that do not Necessitate a High Degree of Detail, such as Tree Counts and High-altitude Drone Flights, will Produce Adequate Resolution and Reduce Flight Duration, Hence Enabling Broader area Aoverage

Drone Altitude for Crop Analysis or Plant Phenotyping (High Detail)

These crop assessments necessitate the highest degree of detail and photographic quality. GSD should be reduced to less than 2 cm and ideally less than 1 cm, contingent upon the specific measure, crop type, and growth stage.

- A general guideline for calculating target Ground Sample Distance (GSD) is to utilize a straightforward calculation.
- Target GSD = Average Plant Size ÷ 30 px
- Employing this algorithm guarantees that each plant is depicted by a minimum of 30 pixels in the images, which typically provides an adequate degree of detail for AI-driven analyses such as plant enumeration.
- The majority of drones must operate at altitudes between 15 and 30 meters. Recent cameras, including panchromatic sensors, can achieve this level of detail at elevated altitudes.
- GSD objective: <2 cm
- Elevation: 15-30 meters (50-100 feet)
- Prevalent Applications:
    - Assessing emergence rates
    - Plant population assessments throughout the initial growth phases of crops such as corn, soybean, and sugar beets.
    - Determining plant dimensions, such as in pre-harvest vegetables like cabbage
    - Plant density/voids
    - Height of the plant
    - Characterizing cultivars (inflorescence, foliar morphology/color, fruiting)
    - Vegetable cultivation experiments utilizing plant-specific analytics

Figure: 5.5 Low-Altitude Flights Facilitate Plant- or Leaf-level Analytics, But Incur Costs Related to Extended Flight Duration and Data Processing.

Adapting to Conditions

Each agricultural and drone operation is distinct. Weather, season, lighting, soil variation, plant species, and additional factors will influence your GSD and flying altitude decision-making.

Utilize our optimal practices as a foundation and consistently tailor them to your requirements and situational factors. Do you still have inquiries regarding altitude?

Drones Simplifying Agricultural Drone Imagery

1. Drones ought to accelerate and improve agricultural management and research. If your drone program is hindered by subpar imagery, we can assist.
2. Solvi streamlines the analysis of agricultural drone footage, allowing growers and academics to concentrate on crops rather than programming.
3. However, simplicity does not equate to superficiality. Our cloud-based analytics platform enables in-depth exploration of crop analytics, including precise plant counts and dimensions, vegetation indices for assessing plant health, biomass, fertility response, plant stress, and additional metrics.
4. Solvi provides comprehensive agricultural drone picture stitching, analysis, and reporting for enhanced crop insights. Experiment with your field images in Solvi at no cost today.

AGRICULTURE DRONE COVER

Agricultural drones have transformed contemporary farming by facilitating precision agriculture methods, improving crop yields, and diminishing the necessity for manual labor. For farmers contemplating the use of these drones, comprehending their coverage capacity is essential. Several critical aspects are involved, including the drone's technological specifications and the attributes of the ground. Let us examine these aspects, their influence on drone efficiency, and methods to optimize your drone's capabilities.

Key Factors That Affect Drone Coverage

The extent of coverage of agricultural drones is determined by various criteria, including the drone model, battery longevity, payload capacity, crop variety, topography, and mission type. Each of these factors significantly influences the efficiency and overall coverage of your drone.

Drone Model

Not all agricultural drones are equivalent. Certain models are intended for small farms, but others might encompass extensive areas of land. Larger drones generally provide superior coverage capacity for agricultural applications because to improved battery longevity, increased payload capacity, and sophisticated navigation systems.

Battery Life

The longevity of the battery is a crucial determinant of a drone's operational range. Drones with extended battery life can remain aloft for longer durations, thereby encompassing a greater area per flight. Battery life is influenced by variables including drone velocity, meteorological conditions, and mission classification. Effective power management during intensive operations such as spraying or sowing directly influences the range of agricultural drone coverage.

Payload Capacity

Payload capacity is essential for tasks such as pesticide spraying, herbicide application, or sowing. A drone with an increased cargo capacity can cover a more extensive region prior it need a refill. Nevertheless, transporting a bigger load frequently diminishes battery longevity, necessitating a compromise between material capacity and flight time. This trade-off affects the coverage area of agricultural drones.

Crop Type and Terrain

The crop variety and topography can also affect coverage. Dense crops such as wheat or maize may necessitate more meticulous spraying, hence reducing the drone's flight speed. Conversely, open fields may facilitate swifter flights and wider spray coverage. Irregular or undulating terrain might impede a drone's navigation systems, impacting the operational range and coverage of agricultural drones. Adapting to diverse terrains is essential for optimizing drone utilization across various environments.

Mission Type

The particular mission influences the drone's operational range. Diverse functions, including spraying, aerial imagery capture, and seed planting, impose distinct requirements on battery life, velocity, and payload capacity. For instance, agricultural spraying necessitates uniform velocity and force, whereas aerial imaging can be obtained from greater elevations with reduced battery consumption. The mission type determines the effective operational range of agricultural drones.

STRATEGIES TO ENHANCE THE COVERAGE AND EFFICIENCY OF YOUR AGRICULTURAL DRONE

Comprehending the elements that restrict coverage is merely the beginning. To enhance your drone's range and efficacy, contemplate these strategies:

Optimize Flight Paths

Formulating optimal flight trajectories is essential. Minimize redundant overlap among routes to optimize time and energy efficiency. Utilize sophisticated tools to compute optimal routes, maximizing coverage while minimizing time and assuring even distribution.

Utilize Battery Management

Effective battery management is crucial for optimizing coverage range. Maintain charged extra batteries and contemplate drones that facilitate mid-flight battery exchanges. Assessing battery performance and modifying flight speed or mission duration might further improve coverage.

Calculating Coverage: The Formula

Although formulas yield approximate estimations, engaging with specialists such as Terra Agri guarantees more precise outcomes. Utilize the subsequent formula for an initial estimation:

Coverage Area = Drone Velocity x Flight Duration x Spray Width

- **Drone Velocity:** The rate of flight of the drone.
- **Flight Duration:** The drone's operational battery life during flight.
- **Spray Width:** The extent of the spray pattern or the region included each application.

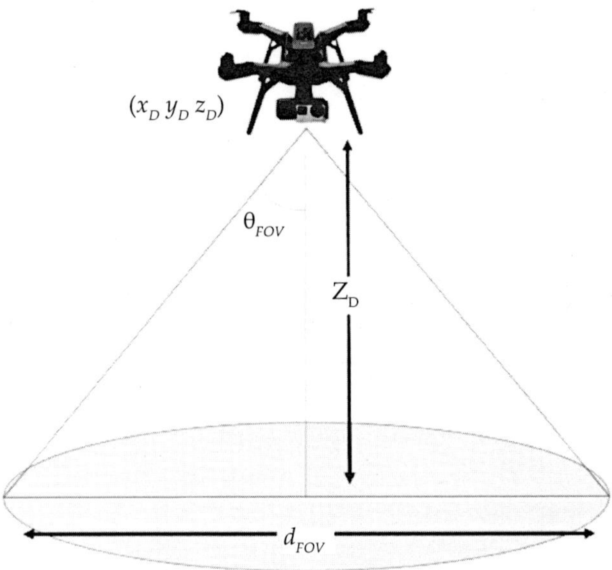

Figure: 5.6 Graphical Concept of the Device Field of View

This formula customizes computations according to your drone model and mission type. Nevertheless, expert counsel from firms such as Terra Agri can offer more accurate direction. Terra Agri provides technical assistance in increasing the coverage capacity and enhancing the effectiveness of agricultural drones, leveraging their diverse variety of drones and considerable experience.

Ensuring the Correct Overlap

Achieving the appropriate overlap of drone pictures is crucial for effective data processing. Aerobotics necessitates a 75% overlap and sidelap, configured via the Flight Planner App. Nonetheless, this overlap is contingent upon your departure point, making the location of your takeoff significant.

- Please refrain from detaching from the lower section of your assignment. This will minimize overlap between photographs where the terrain exceeds the takeoff elevation.
- Ascend from a location proximate to the apex of your quest. This will augment overlap as the terrain descends below the takeoff elevation.
- Ensure that each flight encompasses a complete boundary, avoiding the capture of a partial boundary across many flights. While not always feasible, please endeavor to circumvent this.

A flight must not be conducted if the height gain exceeds 30 meters. The elevation above ground must not surpass 120 meters. Consequently, if there is a 30m elevation disparity, you should ascend from the highest point and adjust the altitude to 80/90m. If you are concerned about reduced resolution imagery when traversing steep slopes, divide the mission into distinct flights with varying departure locations.

When navigating locations with steep inclines, it is advisable to fly parallel to the slope's contour.

Setting the Flight Parameters

The flight settings are primarily determined by the species being mapped and the necessary image resolution. We propose the following recommendations:

- **Altitude:** For orchards, it is advisable to fly at an elevation exceeding 150m/500ft (ensure compliance with local air regulations); conversely, for mapping field crops in their initial stages, a flight altitude of 50m/165ft is recommended. The minimum plant height for Agremo analyses is 12 cm (5 in).

  Utilize 60% sidelap and 70% front lap for orchards, and 70% sidelap and 80% front lap for early-stage field crops.
- **Low-Resolution Example:** Altitude 150m/500ft; Sidelap 60%; Frontlap 70% = 5cm/pixel; 2in/pixel
- **High-Resolution Specification:** Altitude 50m/165ft; Sidelap 75%; Frontlap 80% = 1.5cm/pix; 0.6in/pix
- **Recommendation:** Reducing altitude and increasing overlap will enhance the quantity of photographs and extend flight duration; however, this typically yields high-resolution images.
- Altitude (AGL = above ground level) denotes the height of the flight in relation to the launch site of the drone.
- Overlapping denotes the proportion of both side lap and front lap. The side lap denotes the percentage of overlap between each flight leg, whereas the front lap indicates the percentage of overlap between consecutive images.
- Altitude and overlap are indeed critical flight planning criteria; however, numerous more aspects must also be considered.

The optimal period for agricultural mapping is between 10 AM and 2 PM, as the sun casts minimal shadows during this time. Overcast conditions are generally fine; however, ensure the absence of significant cloud gaps, since they may result in undesirable shadows in your footage. It is generally preferable to do mapping when the entire area is overcast rather than postponing until some clouds dissipate, resulting in a few clear patches.

- The optimal flying speed is contingent upon the desired image resolution. The flight planning application determines flying speed depending on altitude and overlap.
- Modifying the flight route will impact both your flight duration and the quality of the map; ensure that the selected route satisfies the necessary standards.
- Polygon delineation: prior to mission planning, ensure all necessary polygons are designated. Recommendation: the polygons should exceed the dimensions of the field you intend to map.
- Additionally, due to the inability to map extensive fields in a single trip, it is essential to establish mapping zones. Ensure that the overlap in these zones is sufficiently substantial.
- **Recommendation:** Individual zones should not exceed dimensions of 500×500 meters or 1650×1650 feet. One legal requirement for drone operators is that the aircraft must remain inside Visual Line of Sight (VLOS).

BEST PRACTICES FOR EFFECTIVE DRONE FLIGHT PLANNING

Conduct a Pre-Flight Survey

Figure: 5.7 Examine the Place in Advance to Discern Possible Impediments, Including Trees, Electricity Wires, or Structures. This Phase is Essential for Formulating Secure and Effective Flying Trajectories.

Leverage Advanced Software

Employing sophisticated drone flight planning software guarantees fast route formulation, precise overlap, and seamless data integration. Optimal platforms provide user-friendly instruments for the effortless creation and management of missions.

Optimize Flight Patterns for the Task

Select appropriate drone flying patterns according to the project requirements. For instance, grid patterns are effective for mapping extensive areas, whereas linear patterns are appropriate for pipeline inspections.

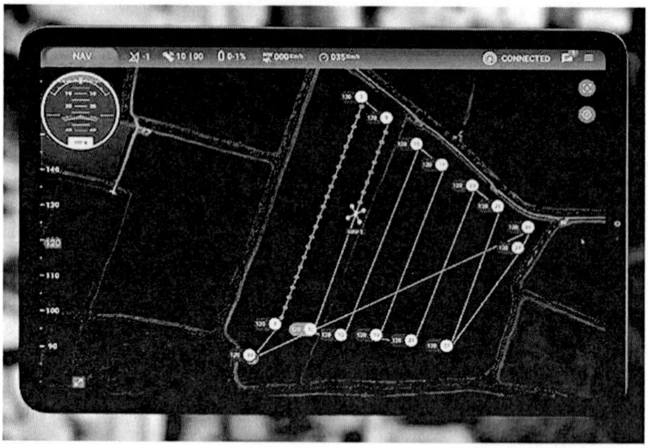

Figure: 5.8 Optimize Flight Patterns

Monitor Weather Conditions

Figure: 5.9 Wind, Precipitation, and Illumination Conditions Substantially Influence Drone Efficacy and Data Integrity. Schedule Flights Within Optimal Weather Conditions for Superior Outcomes

Validate and Test the Mission

Prior to completing the complete mission, do a test flight to verify that the parameters and equipment operate as anticipated.

Figure: 5.10 Test Flight

USE OF MISSION PLANNING SOFTWARE AND TOOLS

How does Drone Mapping Software Help with Agricultural Planning

Drone mapping software for agriculture utilizes a methodical approach to acquire, process, and analyze drone data, facilitating thorough mapping of agricultural terrains. This is a concise overview of the software's functionality.

Figure: 5.11 Drone Mapping Software

Mission Planning

- **Area Definition:** Farmers or agronomists delineate the region they intend to map utilizing the drone mapping software. This may encompass particular areas or sectors of interest within their agricultural land.
- **Flight Path Design:** The program facilitates the design of the drone's flight trajectory. It establishes the best pathways and elevation for the drone to acquire high-resolution photographs with sufficient overlap, guaranteeing precise data collection.

Drone Image Capture

- **Flight Execution:** The drone, outfitted with requisite sensors and cameras, adheres to the predetermined flight trajectory while capturing a sequence of overlapping photographs of the agricultural region. These photographs are captured from diverse angles and perspectives to comprehensively encompass the entire location.
- **Multispectral or RGB Imagery:** Drones can take multispectral or RGB (Red, Green, Blue) imagery, offering several data types that might elucidate different facets of crop health, soil conditions, or other agricultural characteristics.

Data Processing

- **Image Stitching:** After the drone acquires the images, the drone mapping software processes them using photogrammetry techniques. It aligns, corresponds, and integrates the overlapping pictures to provide a unified and precise map.
- **3D Modeling:** The software generates intricate 3D models of the agricultural environment from the stitched pictures. These models offer a three-dimensional depiction of the ground, facilitating a more thorough comprehension of the region.

Analysis and Interpretation

- **Crop Health Assessment:** The software evaluates collected data to examine plant health, growth patterns, and potential issues including pests, diseases, or stress affecting the crops. This analysis is enabled by sophisticated algorithms and sensors that interpret the acquired picture.
- **Decision Support:** Decision Support: The produced maps, models, and analytical outcomes provide significant insights to farmers. They can make informed decisions about irrigation schedules, fertilizer use, pest control tactics, and comprehensive crop management plans.

Reporting and Action

- **Visualization and Reporting:** Drone mapping software displays the analyzed data in accessible visual formats, such as maps, charts, and graphs. These visual representations facilitate farmers' understanding and response to the collected information.

- **Actionable Insights:** Equipped with these insights, farmers may undertake timely and focused measures, executing precise interventions to enhance crop health, augment production, and elevate overall agricultural efficiency.

Drone mapping software for agriculture integrates data from drone-acquired photos, interprets it using advanced algorithms, and delivers actionable insights that enable farmers to make informed decisions for improved productivity and sustainable farming practices.

Role of Software in Drone Flight Planning

Contemporary drone flight planning software has transformed the preparation and execution of missions. These tools not only facilitate crucial procedures and optimize flight routes but also manage data post-collection via drones, encompassing storage, processing, visualization, analysis, and reporting.

Key Functionalities Include

500pxmated Flight Path Generation: The software computes the most efficient flight trajectory according to mission specifications, guaranteeing maximal coverage.

- **Real-Time Monitoring:** Operators can oversee the drone's trajectory and implement modifications during the flight.
- **Data Integration:** Acquired data is promptly transmitted to the cloud for processing and analysis.

Premier flight planning software enables users to optimize mission preparation, oversee operations in real time, and analyze high-resolution data for relevant insights.

The Importance of Post-Flight Analysis

Flight planning concludes only upon the drone's landing. Post-flight analysis guarantees that the obtained data aligns with the mission's objectives. The flight planning program you are utilizing need to streamline this procedure by offering a cohesive interface for data processing, analysis, and visualization.

Essential components of post-flight analysis encompass:

- **Image Stitching:** Transforming photos into precise orthomosaics and three-dimensional models.
- **Data Validation:** Verifying the absence of gaps or discrepancies in the recorded data.
- **Insight Generation:** Sophisticated analytical instruments transform raw data into actionable insights to enhance decision-making.

How the Right Software Enhances Drone Flight Planning for Mapping and Surveying?

The selected software must be a robust platform that facilitates all phases of drone flight planning and data processing.

Intuitive Mission Planning

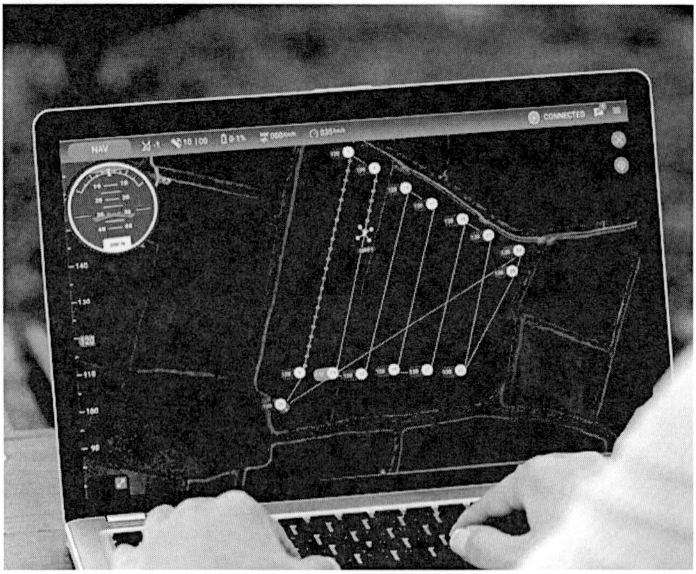

Figure: 5.12 The Interface Must be Intuitive, Facilitating the Creation and Management of Flight Missions while Assuring Optimal Path Generation and Parameter Tuning

Scalability Across Projects

It must accommodate diverse project needs for both single-site surveys and multi-site operations, guaranteeing consistency and efficiency.

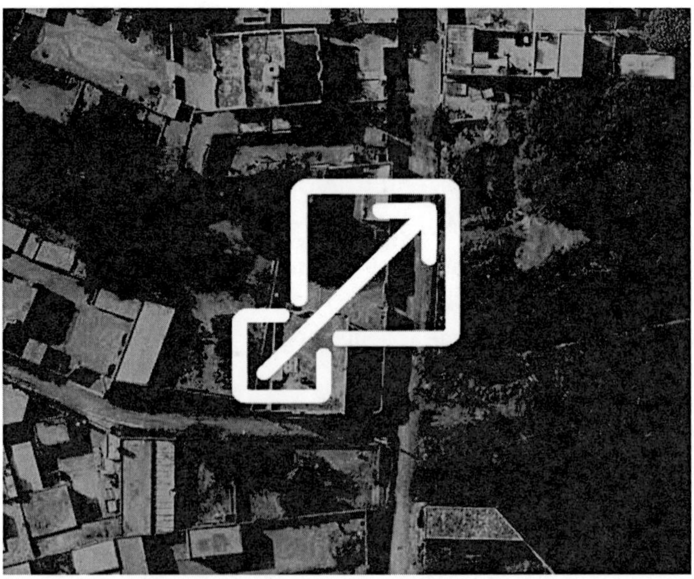

Figure: 5.13 Scalability Across Projects

PRE-FLIGHT CHECKS AND SAFETY PROTOCOLS

Future Trends in Drone Flight Planning

AI-Driven 500pxmation

Artificial intelligence will be important in automating flight planning, optimizing routes based on real-time data, and minimizing manual intervention.

Integration with 5G

Improved communication via 5G networks will provide real-time modifications and data transmission during operations.

Edge Computing

Onboard computing enables drones to examine data during flight, diminishing reliance on post-flight analysis.

Environmental Adaptability

Advanced algorithms will consider dynamic environmental parameters, including as wind and terrain alterations, to ensure mission success under diverse scenarios.

Efficient drone flight planning is fundamental to effective mapping and surveying operations. Utilizing modern solutions such as SkyDeck enables operators to guarantee precise data collection, efficient workflows, and actionable insights. SkyDeck equips customers with the essential capabilities for excellence in geospatial operations, from mission planning through to post-flight analysis.

As the drone business advances, technologies such as SkyDeck will lead, allowing professionals to fully utilize drone technology for mapping and surveying. Meticulous flight planning utilizing appropriate software is essential for success in doing GIS field surveys, generating 3D mappings with drones, or checking infrastructure.

6

Techniques for Data Acquisition During Drone Flights

The drones are configured to traverse the field and acquire photos and sensor data in alignment with the predetermined flight plan. High-resolution cameras get visual data, whereas multispectral sensors gather information on light wavelengths beyond the visible spectrum, thereby offering insights into plant health and stress levels.

During drone flights, essential techniques for data acquisition encompass photogrammetry utilizing overlapping high-resolution images, GPS tracking for accurate location data, IMU (Inertial Measurement Unit) sensors for monitoring attitude and movement, ground control points (GCPs) for precise georeferencing, multispectral or hyperspectral imaging for comprehensive spectral analysis, LiDAR scanning for the generation of 3D point clouds, and data logging systems to document sensor readings throughout the flight, all while meticulously planning flight paths and accounting for variables such as lighting and weather conditions.

Drones have emerged as a vital competitive advantage in the inspection of various asset types, optimizing time, reducing costs, and mitigating risk. Historically, an inspection necessitated an inspector to ascend the structure, capturing images and doing measurements manually from considerable heights. This procedure is intricate; it necessitates extended hours and the presence of numerous experts on site for safety. The data generated by inspectors through manual pictures and measurements is limited: manual inspections may necessitate many trips to collect adequate information for necessary maintenance.

Manually operating a drone to obtain the data necessitates training, expertise, and extensive experience. A pilot collecting data for a tower inspection must comprehend the requirements of the inspectors. The pilot must comprehend how to maneuver an aircraft accurately around an asset to obtain precise aerial photos, regardless of barriers, wind conditions, or the necessity to replace batteries during the trip. The data must be examined and converted into a digital format that is beneficial to the client; identifying potential issues from hundreds of photographs and creating an accurate asset replica can be a labor-intensive operation necessitating specialized expertise.

COLLECTION OF DATA THROUGH DRONES

Drones are unmanned aerial vehicles (UAVs) operated and controlled without a human pilot onboard; nevertheless, they encompass more than mere flying robots.

1. Drones are a vital and versatile instrument for data acquisition. They gather substantial quantities of precise, high-resolution data in a more secure manner.
2. Nevertheless, several types provide drone autonomy. Although the drone is capable of autonomous flight, it necessitates human intervention to either input a predetermined route and coordinates or to activate a Follow-me mode.
3. Drones' data capturing platforms have transformed information gathering across numerous industries. Drones now enable organizations to capture photographs, conduct surveys, create maps, record videos, and perform asset data digitization in inaccessible locations.
4. Drone inspection data management can benefit a firm in various ways. It is capable of performing an airborne inspection, processing the imagery, and subsequently uploading it to a cloud-based platform. Construction sites can utilize automated drone inspections to verify safety compliance and additional factors.
5. Diverse sectors can utilize drone data to obtain valuable business insights. Numerous applications exist for drones, with the predominant usage being to facilitate a company's digital transformation.
6. Autonomous drone devices eliminate numerous variables and challenges associated with airborne data collection. Drone autonomy enables customers to obtain necessary findings without the complexity of gathering insightful data.

A platform for autonomous drone inspections does many of the challenging tasks on your behalf," states Tomer. The software anticipates your data requirements, optimizes the flight path accordingly, and seamlessly resumes the survey following a battery replacement. Machine-learning software can analyze data, pinpointing industry-specific issues and employing photos to construct accurate 3D models of industrial structures.

The Benefits of Drone Data Capture

Accurate data collected during autonomous drone inspections enables enterprise clients to generate precise digital reproductions of their assets, referred to as Digital Twins, facilitating seamless data exchange across extensive enterprises. This is essential for enabling enterprise companies to align their existing historical data in asset management systems with current visual evidence of the precise contents of each asset.

Digital twins can additionally facilitate functions within the organization, including quality assurance, engineering, sales, and security.

The Advantages of Using Drones for Data Capture

- **Enhancing safety:** A primary benefit of utilizing drones is the elimination of individuals from hazardous and perilous settings. Rather than deploying people to hazardous situations, a drone can be utilized to inspect and scan these areas remotely, significantly enhancing safety and reducing danger.
- **Enhanced accessibility:** The dimensions and mass of drones enable them to reach locations that are not easily accessible to individuals. Drones obtain reliable data and imagery at closer proximity and with enhanced resolution.
- **Progress monitoring easily:** Drones facilitate frequent data acquisition and progress assessment of assets.
- **Workflow integration:** Economical, commercially available drones provide substantial value relative to their cost. Incorporating drones into existing workflows enhances data quality and design solutions, hence adding significant value to the project. Mobilizing a crew to collect airborne pictures and data may now be executed with enhanced safety, precision, and broader dissemination.

Defining Agricultural Drone Mapping

Agricultural drone mapping involves the utilization of unmanned aerial vehicles (UAVs), known as drones, outfitted with diverse sensors and imaging technology to gather essential data regarding crops, soil, and terrain. The gathered data is further processed and analyzed to yield meaningful insights for farmers and agronomists.

DRONE MAPPING EQUIPMENT AND TECHNOLOGY

Agricultural drone mapping depends on specific apparatus and sensors to collect and evaluate data.

Drone Hardware and Sensors

Agricultural drones employ several aerial sensor systems designed for crop surveillance.

Multi-rotor drones facilitate vertical takeoff and landing, exhibiting exceptional maneuverability, however they have limited flying durations of 15 to 30 minutes. Fixed-wing drones utilize forward propulsion to traverse significantly bigger areas every flight, although necessitate runway-like takeoffs. Hybrid VTOL fixed-wing drones provide the advantages of both configurations.

High-resolution RGB cameras deliver standard viewable imagery for the investigation of color variance in the field via image processing. Multi-spectral sensors detect narrow non-visible wavelengths that correspond with indications of plant health, including chlorophyll levels. Thermal cameras assist in detecting regions with inadequate irrigation and water stress.

LiDAR laser scanning generates 3D terrain models that assist in identifying drainage issues and facilitate accurate variable-rate irrigation. Hyperspectral sensors offer comprehensive study of chemical composition to assess nutrient status.

Data Analysis Techniques

Specialized algorithms and procedures analyze the aerial imagery obtained from the drones:

- Image segmentation categorizes pictures into regions according to color and crop vitality, facilitating the development of zone-prescription maps. This facilitates targeted interventions, such as irrigating or fertilizing solely the insufficient regions.
- Object identification methodologies can autonomously quantify crops, analyze yields, and delineate weeds and trees with greater efficiency than manual visual evaluation.
- Statistical classifiers that evaluate spectral bands across various wavelengths can forecast critical characteristics, including plant stress levels, yield estimates, and nutritional deficiencies, facilitating proactive early intervention.

The integration of computer vision and artificial intelligence enables drone data to yield comprehensive, actionable insights regarding crop condition and health.

Data Visualization

The GIS-based agronomic analysis program visualizes drone-acquired data via prescription maps, analytical zone delineations, time-series animations, and comprehensive crop analysis dashboards. Interactive graphical interfaces offer farmers and agronomists relevant knowledge regarding field variability and crop performance to aid decision-making.

APPLICATIONS OF AGRICULTURAL DRONE MAPPING

Among the numerous applications that drones offer to farmers, some of the more notable include:

Crop Health Monitoring

Analyzed multispectral drone data utilizing vegetation indices identifies the emergence of pests, illnesses, and nutrient deficits via color alterations prior to the manifestation of obvious symptoms. This facilitates prompt intervention to safeguard yields. Temporal analysis monitors agricultural growth patterns for production forecasting.

Soil and Irrigation Management

Drone-generated soil maps demonstrate moisture variability that suggests inadequate drainage, whereas thermal data identifies irrigation inaccuracies. Improvements in variable rate irrigation and drainage mitigate these concerns. Maps of soil organic matter inform fertilizing practices. Three-dimensional surface models facilitate accurate land grading for enhanced irrigation efficiency.

Land Use Planning and Environmental Impact

Land use analysis based on crop maps, microclimate modeling, and drone observation of wildlife patterns helps plan farms that combine productivity and sustainability. Analysis of erosion, vegetation loss, and water quality parameters offers continual assessment of environmental impact and suggestions for mitigation.

Data Integration with Farm Management Systems

The seamless integration of drone mapping data with modern farm management systems heralds a new era in precision agriculture:

Farm Management Software

Analytics from drone mapping are connected through APIs with existing farm management software solutions to align with operational processes for planting, fertilizing, irrigating, and other field activities. This provides a central platform for operational decision-making by combining drone data with equipment, crop, weather, and other farm data.

Real-Time Data Access

Cloud-based drone data platforms allow farmers and agronomists rapid access to drone-captured maps and analytics through mobile apps. This on-site data accessibility coupled with notifications enables real-time modifications like altering irrigation levels based on drone results. As mapping frequencies grow, data-driven farm course corrections become faster.

Benefits of Agricultural Drone Mapping

Growing healthy, thriving crops needs critical observation, understanding, and proactive care. But even the most vigilant farmer can overlook early indicators of pests, illness, or inadequacies disguised amid acres of green foliage. Utilizing agricultural drone mapping, farmers can obtain insights by monitoring the plants, among other factors. Agricultural drones offer numerous advantages:

Sharpening the Senses : Drone Data Reveals Crop Truths

Drone mapping offers an advanced airborne viewpoint for evaluating crop health and soil conditions. On-board cameras can assess field fluctuations with centimeter precision, uncovering insights imperceptible to the naked eye. Spectral sensors identify minor physiological alterations in plants by assessing light absorption and recognizing the initial signs of stress. Thermal cameras delineate underlying moisture, facilitating accurate irrigation. Three-dimensional soil topology models detect drainage issues. Drones unveil the authentic narrative of each plant through data that would necessitate exhaustive walking and manual measurement. This plant-centric approach facilitates precise interventions to cultivate robust crops.

Maximizing Yields Through Prescriptive Action

Drones facilitate the judicious allocation of resources by identifying particular field requirements. Drone-generated nutrient maps provide variable rate fertilizing according to soil nutrient concentrations. Plant health photography facilitates targeted pesticide application exclusively to impacted regions, hence minimizing chemical usage. Analysis of growth patterns helps forecast yields weeks ahead, facilitating the planning of harvest logistics and storage. Drones enhance productivity and minimize waste through the facilitation of prescriptive action. The future of agriculture will depend on data-driven, evidence-based crop recommendations.

Vigilance From Above : Catching Threats Early

Crop hazards, including pests, fungus, and bacteria, can proliferate swiftly and markedly diminish production and quality if not identified promptly. Drones provide ongoing surveillance to identify issues promptly, preventing their escalation into infestations. Subtle alterations in vegetal coloration, imperceptible to the naked eye, can be identified with spectral cameras. This preliminary knowledge facilitates the implementation of containment strategies prior to the proliferation of threats. Drones evaluate the efficacy of administered therapies, facilitating prompt modifications. Aerial surveillance is crucial for safeguarding against roughly 63,000 types of agricultural pests globally.

In summary, drone crop mapping offers farmers prompt, comprehensive crop information unattainable through ground observation alone. This facilitates precise actions to enhance inputs and safeguard crops. The future of agriculture will leverage data from novel aerial perspectives to develop the healthiest and most productive farming possible. Drones have become essential instruments for contemporary, data-oriented agriculture.

Challenges and Concerns in Agricultural Drone Mapping

The obstacles along this road require resolution:

Privacy

Proprietary drone-captured agricultural data in cloud systems requires careful security measures and access controls, similar to any data source. Data utilization must adhere to privacy legislation concerning aerial monitoring in proximity to residential zones. Responsible data stewardship preserves farmer trust and safeguards agricultural investments.

Challenges in Technical Matters

Drones are facilitating aerial surveys; yet, restrictions imposed by battery life, weather conditions, flight regulations, sensor capabilities, and data storage capacity necessitate ongoing advancements through study. Interoperability, automation, cyber-physical security, and decision-support functionalities must advance as usage intensifies.

Accessibility and Adoption

The expenses associated with drone hardware, software, and analytical knowledge are obstacles for smallholder farmers' adoption in underdeveloped nations. Customized toolkits designed for small farms may facilitate worldwide adoption. Fostering local service provider ecosystems promotes the distribution of these developing technologies and their advantages.

Data Processing and Analysis

The data is handled with specialist software that amalgamates photos to produce orthomosaic maps (geometrically rectified aerial views) and evaluates sensor data to derive insights. The maps offer a detailed and accurate representation of the area, while the analysis of sensor data clarifies discrepancies in plant health, soil conditions, and other critical elements.

Report Generation and Interpretation

The analyzed data is delivered as reports, maps, and visualizations that are easily understandable for farmers, facilitating informed decision-making. Such assessments may pinpoint areas necessitating attention, such as nutrient deficits or insect infestations, allowing farmers to execute targeted interventions and enhance crop management practices.

POST-FLIGHT DATA PROCESSING AND ANALYSIS

LOGIS offers post-flight data processing, analysis, and mapping services for data remotely acquired by UAVs (Unmanned Aerial Vehicles or Drones).

Drones provide a rapid and economical solution for mapping and surveying endeavors, maintaining accuracy, quality, and resolution.

Applications encompass:

- Precision agriculture (vegetation indices)
- Mining, quarrying, and landfill operations (surveying, contour mapping, volume assessments)
- Geographic Information Systems (aerial imagery, ortho-mosaics, three-dimensional models)
- Environmental and catastrophe management (game counts, expedited disaster effect assessments, and monitoring)

We can arrange flights for your research area, route configuration, or region of interest.

Figure: 6.1 Post Flight Data Processing

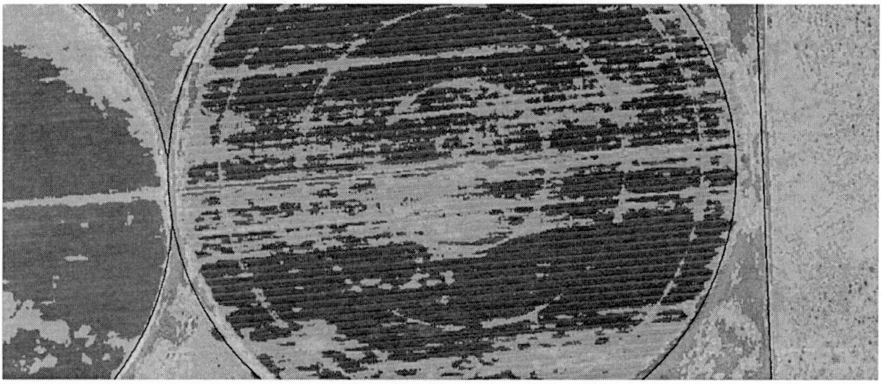

Figure: 6.2 Flight Data Processing

In Field Postflight

Upon completion of its mission, the UAV autonomously returns to and lands at its original takeoff location. Upon landing, extract the memory card from the UAV or sensor and confirm on the laptop that all obtained data adheres to the requisite criteria via a memory card reader. If the data do not meet the acceptable requirements, you must identify which parameter influences the data and restart the task with revised settings.

Common Problems

- **Camera Focus Problems:** Occasionally, the sensor may be misaligned, rendering the data ineffective. You must determine the suggested focus settings for the sensor and repeat the mission. Utilizing manual focus rather than autofocus might occasionally resolve this issue.
- **Data Absence:** Numerous things may contribute to this problem. The primary issue is wireless interference with the remote controller, typically arising from nearby magnetic interference sources or the presence of another remote controller in proximity to the user. A further cause may be communication failure resulting from the overheating of the mobile device. This can be mitigated by positioning the mobile device beneath a shelter or in proximity to an air circulation apparatus. Memory cards may provide a problem if they are incompatible or if they get full during a mission.
- **Overlap Concerns:** The utilized overlap percentages may be insufficient to include the data collection area. This issue typically arises when the sensor operates independently of the UAV. In this scenario, the sensor specifications must be inputted into the flight mission planning application to prevent overlap complications.

Post-Flight Analysis Off the Field

Upon returning from the field with the UAV-acquired data, you must adhere to multiple procedures to generate different sorts of maps. A separate paper provides detailed preflight and flight instructions in a sequential manner.

Collected Data

- Transfer all data from the memory card to the PC in use. A data backup in an alternative location is advisable.
- Examine the gathered data to ascertain its integrity, including the quality and resolution of the photographs.
- When gathering data across multiple fields, categorize and organize the data into distinct folders according to the field numbers.

Ensure the removal of superfluous data. For instance, when a sensor operates independently of the UAV during flight, it begins data collection prior to takeoff and continues until it is deactivated after landing. The data may be irrelevant to the purpose, and their deletion could conserve processing time. An further instance occurs when a single flight services two geographically distant regions. The data captured between the blocks may be redundant and might be eliminated.

Calibration

Nearly all spectrum sensors require calibration. The sensor manufacturer often provides all necessary tools. Adhering to the calibration instructions and procedures is crucial. When utilizing a sensor requiring calibration, ensure you possess the calibration values for each specific band within the spectrum employed. In this instance, it is imperative to gather calibration data prior to a flight mission. This information is typically gathered before to the commencement of the mission, or in certain instances, it constitutes part of the data obtained. The approach employed is contingent upon the sensor's manufacturer. If these data are included in the collected information, the user must locate them manually; otherwise, the data must be gathered prior to the flight. This information will be utilized in post-processing to calibrate the data.

Challenges and Solutions in Drone Data Processing

Data Volume and Storage

- **Challenge:** A significant obstacle in drone data processing is the huge volume of data acquired by drones. High-resolution photographs and movies can rapidly accumulate, leading to terabytes or even petabytes of data. The storage and management of huge data volumes provide considerable obstacles.
- **Solutions:** To resolve this issue, incorporate cloud storage services, employ data compression techniques, and adopt effective data management tactics to prioritize pertinent data for analysis.

Data Transfer and Connectivity

- **Challenge:** The transfer of substantial data quantities from drones to ground stations or remote computers can be a protracted endeavor. Restricted connectivity or remote operational settings can intensify this issue.
- **Solution:** To address this, data transmission protocols must be adjusted to reduce transfer durations. Furthermore, utilizing on-board processing

capabilities in drones might minimize the necessity of uploading raw data, allowing for the transmission of only processed information and so decreasing bandwidth requirements.

Real-time Processing and Analysis

- **Challenge:** Numerous drone applications necessitate real-time data processing and analysis to swiftly derive meaningful insights. Processing extensive datasets in real-time can tax computing resources and lead to delays.
- **Utilizing high:** performance computing systems, parallel processing methods, and edge computing can address these difficulties by dispersing the processing load and facilitating expedited analysis.

Data Accuracy and Quality

- **Challenge:** Guaranteeing the precision and quality of drone data is essential for dependable analysis and decision-making. Variables including illumination, meteorological conditions, and aerodynamics can influence data integrity.
- **Solution:** Implementing image stabilization techniques, calibrating sensors, and employing several sensors for cross-validation can enhance data accuracy. Moreover, instituting quality control protocols during data collection, including onboard sensor diagnostics, can assist in detecting and alleviating any data problems.

Data Integration and Fusion

- **Challenge:** Drone data frequently requires integration and fusion with additional data sources for thorough analysis. This may involve integrating drone photography with satellite imaging or incorporating drone sensor data into geographic information systems (GIS). Data integration challenges may come from discrepancies in data formats, coordinate systems, or resolution levels.
- **Solution:** Implementing data standards protocols, utilizing compatible formats, and employing software solutions specialized for data fusion can enable seamless integration.

Data Privacy and Security

- **Challenge:** Drone data may encompass sensitive information, including private property details or personally identifiable information (PII). Safeguarding data privacy and guaranteeing its security during collection, storage, and processing is paramount.
- **Solution:** Measures such as encrypting data transmissions, enforcing access limits, and complying with privacy legislation can alleviate the risks linked to data privacy and security breaches.

Drone data processing encounters numerous hurdles, including substantial data volume, real-time processing demands, accuracy problems, integration complexities, and privacy considerations. Utilizing technologies like cloud storage, enhanced data transfer protocols, high-performance computing, and stringent data quality control methods can successfully mitigate these difficulties. Surmounting these obstacles will allow businesses and industries to fully leverage drone technology and get important insights for informed decision-making and enhanced operational efficiency.

INTERPRETATION OF AERIAL IMAGERY AND SENSOR DATA

Agricultural drone imagery is obtained by unmanned aerial vehicles (UAVs) outfitted with cameras and sensors to capture high-resolution photos of agricultural land. The drones traverse agricultural areas, capturing intricate photos and data that are subsequently processed and analyzed to yield useful insights for agriculture.

Benefits of Drone Imagery in Agriculture

- **High-Definition Data:** Drones provide very detailed imagery, facilitating accurate assessment of crop vitality and soil conditions.
- **Adaptability and Accuracy:** Farmers can utilize drones to access designated areas and focus on specific sections of land.
- **Temporal Efficiency:** Drones collect data from extensive areas significantly more rapidly than manual approaches.

Use Cases of Agriculture Drone Imaging

- **Aerial Surveillance of Agricultural Fields Using Drones:** Drones furnish real-time information for agriculturalists to oversee crops. Aerial perspectives are utilized for assessing plant health, growth, and indications of stress or sickness.
- **Estimation of Yield:** Drones record discrepancies in crop density, size, and health across various field zones, facilitating more precise predictions of harvest yields.
- **Monitoring of Soil Erosion:** Comprehensive topographic maps and drone imagery disclose erosion patterns and risk zones, enabling farmers to adopt preventive strategies such as contour plowing or cover planting.

Limitations of Drone Imagery use in Agriculture

- **Restriction of Coverage:** Drones exhibit diminished efficacy across extensive terrains due to their limited coverage area. The expense of drone operations is excessively elevated for extensive agricultural regions.
- **Dependence on Weather Conditions:** Adverse weather conditions, such as high winds or rain, might impede drone operations, resulting in delays in data acquisition.
- **Analysis of Data:** Analyzing and interpreting drone imagery necessitates specialized software or expertise.
- **Duration of Battery and Flight Time:** The battery life is limited, necessitating regular recharging or battery replacements for the completion of extensive duties.

SOFTWARE TOOLS FOR DATA PROCESSING AND VISUALIZATION

Drone mapping analytics are connected via APIs with current farm management software to synchronize with operational processes for planting, fertilizing, irrigating, and other agricultural activities. This establishes a centralized platform for operational decision-making by integrating drone data with equipment, crop, weather, and additional agricultural data.

The agriculture drones market is projected to expand from USD 1.2 billion in 2019 to USD 4.8 billion by 2024, reflecting a compound annual growth rate (CAGR) of 31.4%. Pressure and strain on the global food supply are unprecedented due to the rapid growth of the world's population. In contemporary capitalist society, opportunities generally attract investment and growth, exemplified by the advancement of agricultural drones, where significant interest from venture capitalists serves as a primary catalyst for market expansion.

Field mapping constitutes the predominant segment of the precision farming agriculture drones market. Field mapping assesses the condition and vitality of crops across various soil types and management zones to illustrate crop health and yield potential.

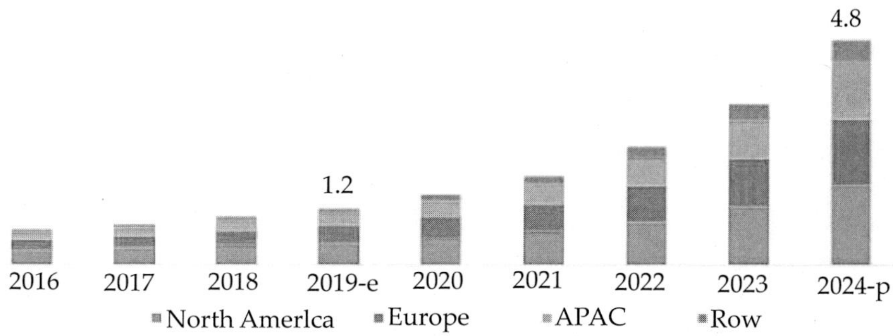

Figure: 6.3 Agriculture Drones Market, by Region (USD Billion)

The agriculture drones market is anticipated to experience a higher compound annual growth rate (CAGR) over the projected period. Significant investments from venture capitalists and angel investors indicate that software and analytics for digitizing data gathered by drones will be a prominent sector in the future. Principal fields for investment include mapping, imaging, and data analytics software.

North America is poised to dominate the agricultural drone market due to the Federal Aviation Administration (FAA) exemption under the Part 107 regulation, resulting in increased drone adoption in agriculture. This fact draws investments into the agricultural drone business.

Drone technology has advanced significantly in recent years, although acquiring aerial imagery is merely the initial step. To fully optimize drone data, appropriate software is essential for processing, analyzing, and transforming raw data into meaningful insights. The selection of tools significantly influences the outcomes of land mapping, infrastructure inspection, and survey execution. This article examines the premier drone data processing software and AI solutions now available, detailing the features of each to assist you in identifying the most suitable option for your requirements.

Pros and Cons of Geospatial Platform

Pros

- Provides an extensive array of AI-powered instruments for geospatial analysis.
- Adaptable pricing structures to meet diverse user requirements.
- Enables users to develop bespoke AI models without requiring programming expertise.

Cons

- Enhanced functionalities and superior processing capabilities are exclusive to premium subscriptions.
- Inadequate support for lower-tier plans, potentially obstructing consumers requiring prompt assistance.

Pros and Cons of Aerial Mapping Solution

Pros

- Robust wind resistance and broad area coverage.
- Versatile camera configurations including straightforward interchangeability procedures.
- Design that is resilient to future developments, accommodating forthcoming AI integrations.

Cons

- Increased first capital outlay relative to standard drones.
- Training is necessary to fully leverage advanced features.

Pros and Cons of Photogrammetry Software

Pros

- Facilitates both cloud and local computing
- Exceptional accuracy in measurements, to the level of a single pixel
- Comprehensive post-processing tools provided
- Compatible with various picture sources, including drones, cameras, and smartphones

Cons

- The SOLO plan restricts cloud processing to a maximum of 150 photographs per project.

- Advanced functionalities and offline processing are exclusively accessible in premium subscriptions.

ADVANTAGES OF FARM MAPPING SOFTWARE

Farm mapping software offers an effective solution that enhances the entire agricultural value chain. Farm mapping systems have fundamentally revolutionized the agricultural sector, enabling farmers to swiftly acquire vital data and convert it into lucrative business decisions. The sustainable management of the complete harvesting life cycle necessitates extensive and meticulous planning. The findings derived from drone-collected data enable precise situational assessment and the implementation of necessary modifications, actions, and enhancements. Furthermore, including a planting database with a map with the collected data enhances future accessibility.

agricultural mapping software is a technology that will elevate agricultural management significantly. The genuine benefits that farm mapping software will provide to your agricultural enterprise are as follows:

Soil and Field Analysis

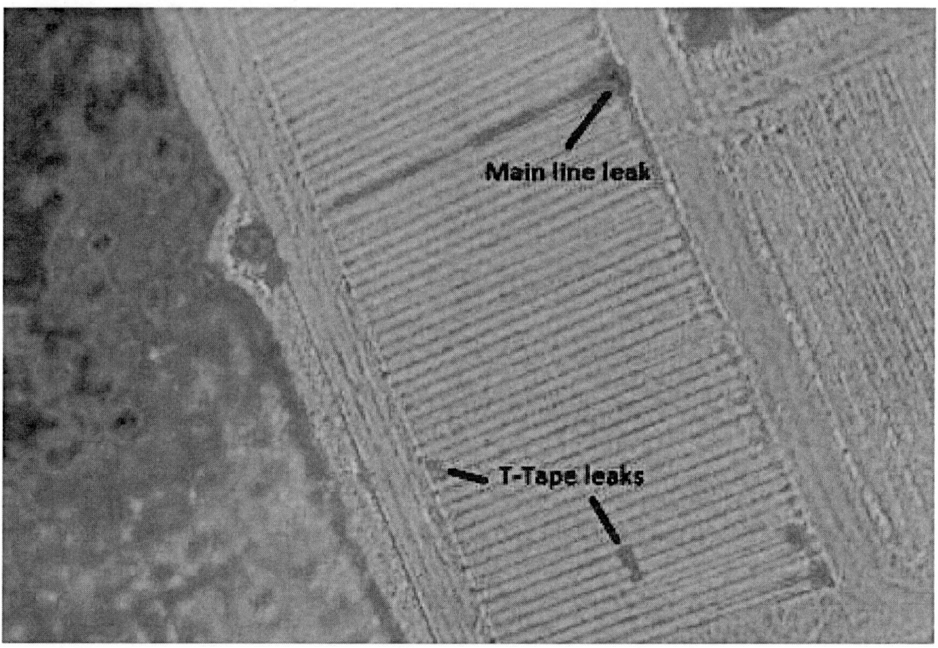

Figure: 6.4 Thermal Imaging Employed to Detect Water Leaks

Planting & Irrigation in AgroTech

Recognizing field conditions enables farmers to make informed decisions, optimize irrigation, and improve nitrogen management. Drones can obtain accurate 3D volumetric data, which identifies elevations and drainage locations, hence reducing the required staff for the task. Farmers can utilize this data to strategize their planting patterns and allocate crops in accordance with landscaping particulars.

Geo-Tagged Monitoring

It must be recognized that extensive field areas significantly hinder or entirely undermine the effectiveness of crop monitoring. However, erratic weather patterns necessitate that farmers act promptly without deferring until the conclusion of a season. It is no longer sufficient or practical to ascertain the causes of failure solely by analyzing aftermaths and seasonal statistics while reflecting on past losses.

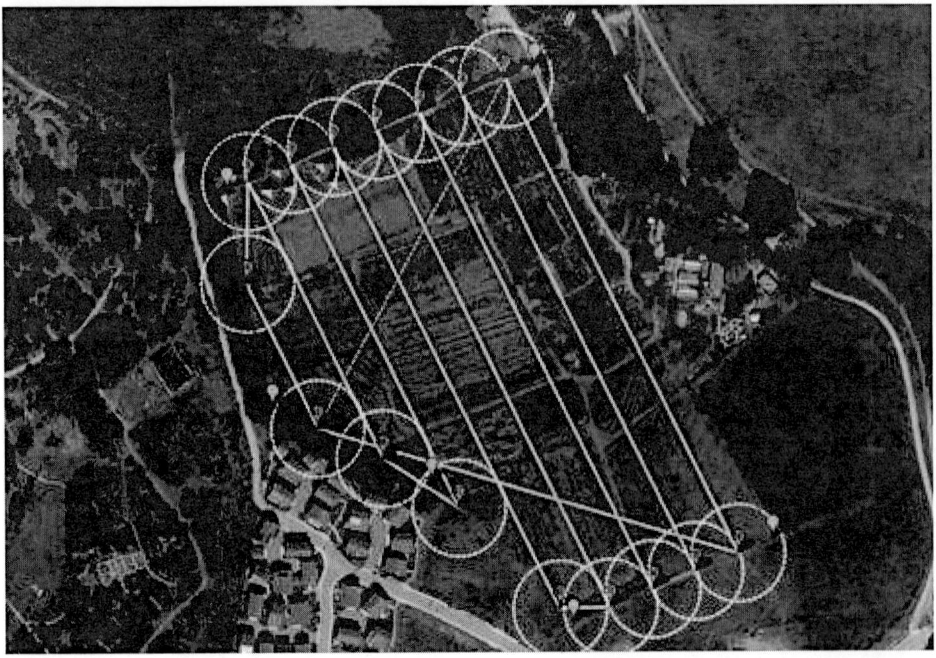

Figure: 6.5 Geo-Tagged Monitoring

Crop Spraying

A professional UAV provides comprehensive insights into crop development, allowing agricultural experts to promptly spot difficulties and enhance their field scouting efforts. Data amassed over multiple years can enhance future planning and oversight, for instance, by providing clearer insights into the optimal placement of ditches and the evolution of fertilizer applications.

For instance, the application of drones throughout various growth stages of corn enhances value throughout the growing season, encompassing emergency inspections, early growth evaluations, and pre-harvest predictions.

Crop monitoring offers adaptability, enabling farmers to modify their strategies during the season and even within a month. Crop monitoring enables farmers to minimize field maintenance expenses and concentrate just on the most vital areas, therefore conserving their time and money.

Historically, crop monitoring was solely achievable through satellite images. This service was evidently expensive, image quality could be compromised by weather conditions, and the resolution was lacking. Fortunately for the sector, Unmanned Aerial Vehicles are poised to remain, evolve, and serve as a superior alternative to their satellite imagery forebears in the future.

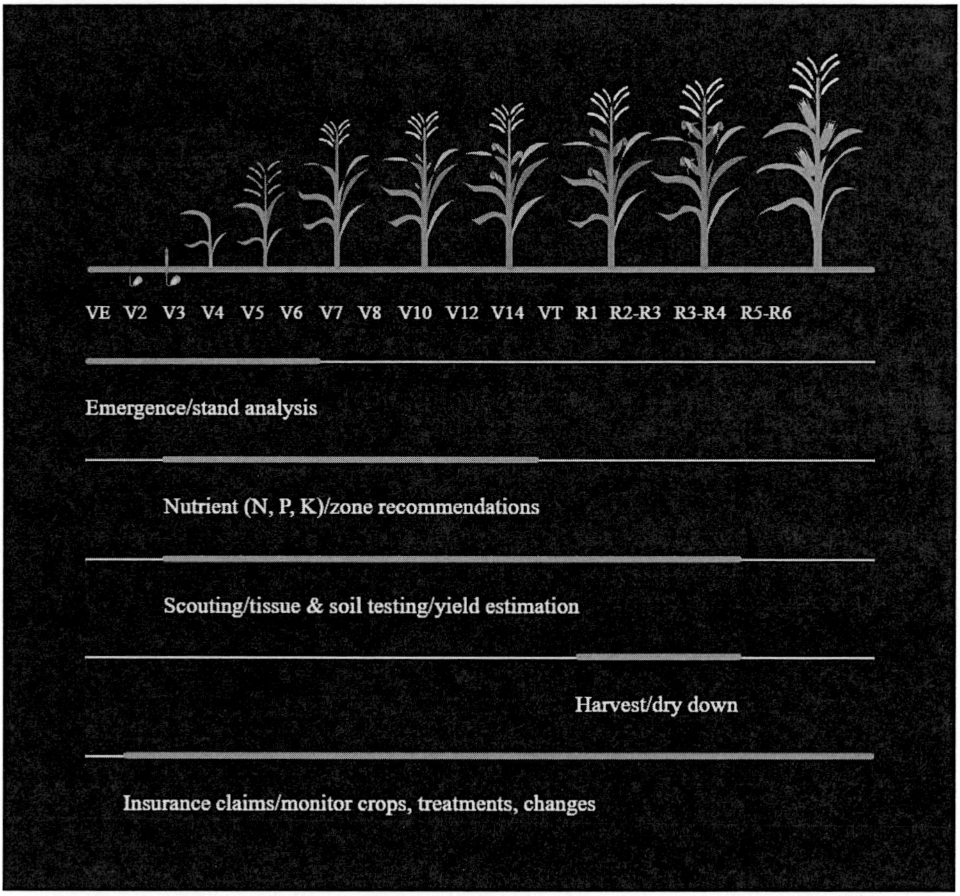

Figure: 6.6 Crop Spraying

Agricultural runoff from industrial fertilizers and pesticides poses a significant health risk. Drones assess plant health and identify problematic regions, generating useful data that enables farmers to minimize their chemical usage by applying substances in a highly focused and judicious manner.

Health Assessment

The health assessment is conducted by collecting data from multiple sources using thermal, multispectral, and infrared sensors. In this manner, farmers can acquire knowledge on the solar absorption and transpiration rates of their crops. Additionally, the NDVI mapping technique can be utilized to assess the health status of the field.

Figure: 6.7 Health Assessment of Farms by Drone

The Normalized Difference Vegetation Index (NDVI) is a mapping technique that identifies green areas on a map and provides early warnings based on subtle color variations in crops.

NDVI is widely recognized among industry professionals. The distinguishing factor of this technology is the superior precision in image quality and acquisition speed attainable with drones compared to satellites. Commercial drones can generate highly detailed maps and measure areas with precision down to inches and centimeters. Drones are superior and more cost-effective than satellites.

With this information available in advance, farmers may potentially achieve not $3,000 to $4,000 per ton, but rather $5,000 per ton, contingent upon the specific issue at hand.

Drone technology can potentially be utilized to detect bacterial or fungal problems in trees. Health information can be obtained from multispectral images that monitor alterations in plants and signify their condition. A prompt reaction can preserve an entire orchard.

Upon the identification of a vegetation illness, farmers can employ remedies to obtain a more accurate understanding and monitor the plant's condition throughout the treatment process to evaluate the efficacy of the administered drug or intervention.

Precision Agriculture

All mapping procedures serve as the foundation for a complete agricultural management concept known as Precision Agriculture. Precision Agriculture has gained significant popularity within the industry and is regarded as one of the most promising trends.

The approach involves observing, measuring, and addressing inter- and intra-field variability in crops. For a more comprehensive perspective, refer to the insightful essay published by The Economist on this subject.

Figure: 6.8 Precision Agriculture

The objective is to allocate the farm's resources more efficiently to achieve optimal yield. This is accomplished by reducing variability in crop health within an individual field. You are likely contemplating the implementation of intelligent software solutions for your agriculture enterprise. Nonetheless, you may find it challenging to determine which solution or solutions will most effectively support your business objectives. You may contact our team to obtain essential answers to all your inquiries. Leveraging years of vast experience in the technology sector and informed by comprehensive business research, Altamira will assist you in selecting and developing optimal solutions for your enterprise.

Mapping procedures serve as the foundation for a complete agricultural management concept known as Precision Agriculture. Precision Agriculture has garnered significant appeal within the sector and is regarded as one of the most promising innovations.

The notion relies on the observation, measurement, and response to variability both within and between crop fields. For a more comprehensive understanding, here is an excellent essay published by The Economist on this subject.

The objective is to allocate the farm's resources more efficiently to achieve optimal yield. This is accomplished by reducing crop health variability within an individual field, hence implementing targeted interventions.

Upon the identification of a vegetation illness, farmers can employ remedies to obtain a more accurate understanding and monitor the plant's condition throughout the treatment process to evaluate the efficacy of the administered drug or intervention.

Agriculture Mapping Software Functions

Precision agriculture necessitates substantial data. Furthermore, it necessitates the regular collection of data. That is where agriculture software proves beneficial.

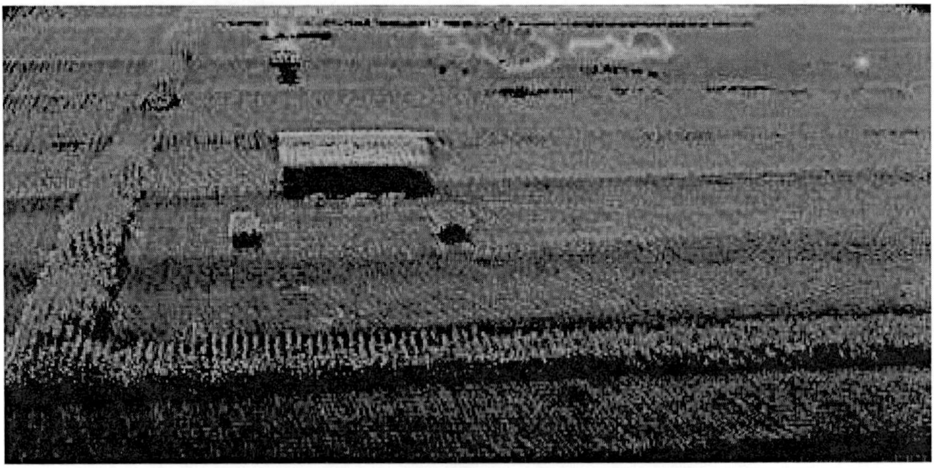

Figure: 6.9 Agriculture Mapping Software Functions

UAV drones are capable of collecting diverse data by capturing images of livestock, fields, and certain plants. Here is a preliminary list of information that can be obtained from images:

- Quantity of plants, height measurement.
- Monitoring of plant stress.
- Leaf Area Indexing.
- The existence of pathogens, nutrients, and weeds.
- Estimation of relative biomass.
- Monitoring of harvests.
- Animal locomotion.
- Classification of trees.
- Evaluation of drought conditions.
- Three-dimensional field surveys (elevations, excavations, optimal water drainage locations).

tions, holes, suitable water drainage points).

How Farm Mapping Software Works

Let's start with the types of imagery that can be acquired by drones:

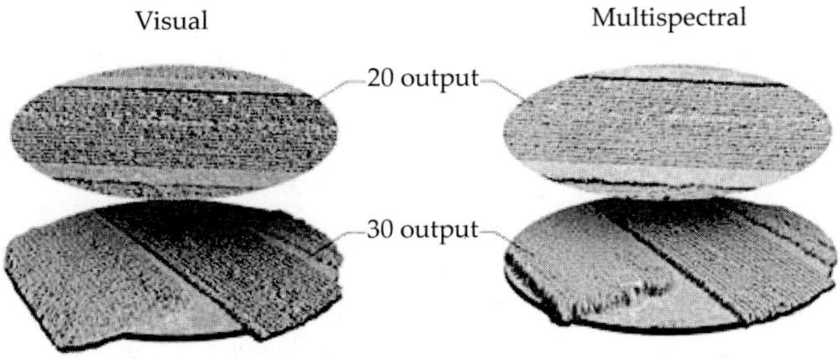

Figure: 6.10 Working of Form Mapping Software

- **Thermal/Infrared:** This type of sensor can detect hotspots and quantify land and vegetation temperature. Thermal sensors are employed to identify water due to its cooler temperature signature, aiding in the detection of drought or irrigation leakage.
- **Multispectral:** Utilized for health evaluation and normalized differential vegetation index.
- **LIDAR Technology:** Notwithstanding its substantial expense, it may be utilized once for preliminary 3D field surveying.
- **Hyperspectral Information:** The amalgamation and superimposition of images from many sources facilitate a more comprehensive examination.

How Agricultural Drones and Software Can Measure Health

Agricultural software predominantly employs multi-spectral images to assess the health of food products at any growing stage. These algorithms examine variations in near-infrared and visible light reflected by crops, utilizing imagery captured by agricultural drones.

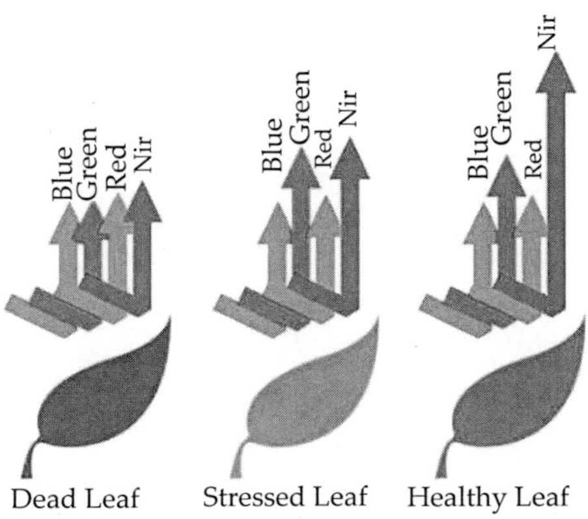

Figure: 6.11 Measurement Health of Food Products by Agricultural Software

The method specifically examines variations in green color volume in VIS images and NIR light. Generally, an increase in green and reflected light correlates with improved plant health.

Farmers can utilize this data to monitor health variations over time by calculating and measuring NDVI. The basis of NDVI calculation is fundamentally based on the same methodology. NDVI = (NIR - VIS) / (NIR + VIS).

This succinctly summarizes the functionality of agricultural health mapping software.

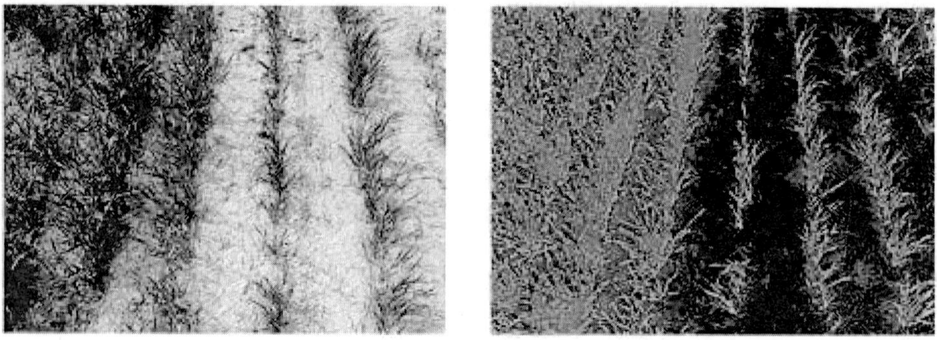

Figure: 6.12 An Example of an NDVI-Processed Image.

7

Applications of Drones in Crop Monitoring

Drones have transformed agricultural operations, especially in crop monitoring, by providing farmers with an efficient and economical method to collect real-time data on their crops. Drones, outfitted with sophisticated sensors, cameras, and imaging technology, can obtain intricate aerial perspectives of extensive agricultural landscapes. This enables farmers to oversee crop health, identify early indicators of diseases, pests, and nutrient deficits, and evaluate water levels or irrigation requirements. Drones offer essential insights for precision agriculture, enabling farmers to make informed decisions that enhance output, minimize resource use, and improve overall crop management. Drones' capacity to swiftly and precisely cover extensive regions has rendered them an essential instrument in contemporary agriculture.

UNDERSTANDING DRONE CROP MONITORING

Crop monitoring using drones is the deployment of drones fitted with specialized sensors and imaging systems to collect data regarding crops and their environment. The data undergoes processing and analysis, yielding outcomes that offer farmers practical insights to enhance crop output, decrease input costs, and mitigate environmental effect. The goal is to employ technology to enhance understanding of crop performance, so enabling more informed management decisions.

The procedure for drone crop monitoring generally encompasses the subsequent steps:

Flight Planning

Establishing the drone's flying trajectory and elevation to guarantee optimal coverage of the designated region is essential. Elements including field dimensions, configuration, and the required data resolution affect flight planning choices. Specialized software frequently facilitates this process by generating optimal aircraft paths that enhance data collecting while reducing flight duration.

Data Acquisition

The drones are configured to traverse the field and acquire photos and sensor data in alignment with the predetermined flight plan. High-resolution cameras get visual data, whereas multispectral sensors gather information on light wavelengths beyond the visible spectrum, thereby offering insights into plant health and stress conditions.

Data Processing and Analysis

The data is handled with specialist software that integrates photos to produce orthomosaic maps (geometrically rectified aerial views) and evaluates sensor data to derive insights. The maps offer a detailed and accurate representation of the field, while the analysis of sensor data clarifies discrepancies in plant health, soil conditions, and other critical elements.

Report Generation and Interpretation

The analyzed data is delivered as reports, maps, and visualizations that are easily understandable for farmers, facilitating informed decision-making. Such assessments may pinpoint areas necessitating attention, such as nutrient deficits or insect infestations, allowing farmers to execute targeted interventions and enhance crop management practices.

Plant Health Assessment

Drone-assisted plant health monitoring is a contemporary and novel methodology in agriculture and land management. It entails employing drones outfitted with diverse sensors and cameras to evaluate the health and status of crops, forests, and other plants. This technology has been popular due to its provision of fast and comprehensive information that assists farmers, land managers, and researchers in making informed decisions on crop management, pest control, and resource allocation. Key components of drone-based plant health monitoring include:

The utilization of drones in agriculture has enhanced crop monitoring by offering farmers an aerial perspective of their fields. Drones are outfitted with sophisticated sensors and high-resolution cameras that document every detail of crops, uncovering possible problems imperceptible to the human eye.

The photographs captured by these drones are utilized to study indicators of stress, nutrient deficits, and early signals of pest or disease outbreaks. Drones, due to their rapidity and efficacy, obviate the necessity for labor-intensive physical inspections, conserving farmers' precious time and resources.

Drones efficiently cover extensive regions in a little period, and the data they collect enables farmers to undertake corrective actions that guarantee optimal crop health and yield.

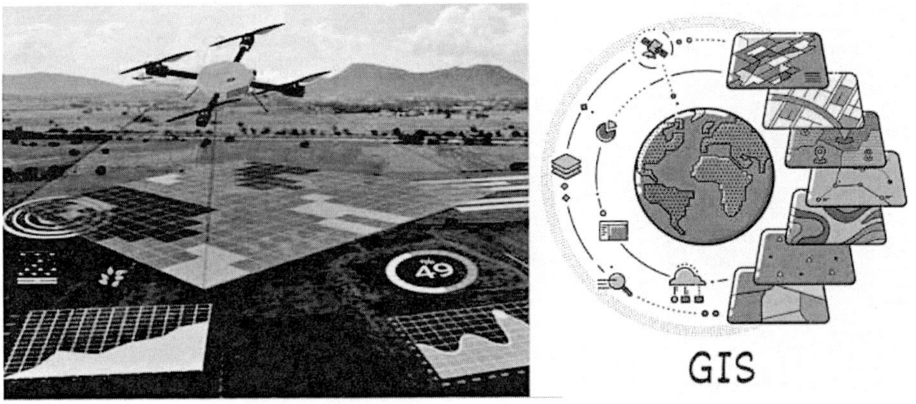

Figure: 7.1 Geospatial Imaging

THE SIGNIFICANCE OF CROP HEALTH MONITORING

Efficient Crop Health Monitoring is crucial for optimizing yields. It also guarantees sustainable agricultural methods. Historically, manual inspections and periodic assessments were integral to crop monitoring. These approaches were arduous and susceptible to human mistake. The advent of drones has transformed the monitoring of crop health. It now provides an enhanced degree of accuracy and efficacy.

How Drones Enhance Crop Health Monitoring

Drones outfitted with sophisticated sensors and imaging technology have transformed the monitoring of crop health. Here is their method of altering the field:

Figure: 7.2 Utilization of Drones for Enhancing Health

Drones acquire high-resolution aerial imagery of complete fields. This offers an extensive overview of agricultural circumstances. This comprehensive picture enables farmers to evaluate plant health over extensive regions. It aids in detecting anomalies and issues that conventional methods may overlook.

- **Multi-Spectral Imaging:** In agriculture, Multi-Spectral Imaging holds significant value. The multispectral drone improves crop health monitoring. It offers high-resolution multispectral imaging capabilities. This drone acquires data across multiple wavelengths. The wavelengths encompass Visible, Near-Infrared, and Red-Edge Bands. This sophisticated sensor system enables farmers to produce comprehensive vegetation indexes. An example of such an index is the NDVI. NDVI signifies vegetation vitality. It accomplishes this by evaluating the light reflection properties of crops. Multi-Spectral Imaging enhances crop vitality by identifying stress and nutrient deficiencies. It additionally quantifies moisture levels that are imperceptible to the unaided eye. Agriculturists can more accurately assess plant health through the analysis of diverse wavelengths. They can also improve irrigation management and economic resources. This results in enhanced agricultural productivity and promotes sustainable farming methodologies. Multi-Spectral Imaging is a significant leap in drone technology. It improves the surveillance of crop health. Drones outfitted with multispectral cameras capture images across several wavelengths. These wavelengths extend beyond the visible spectrum. This skill allows farmers to observe facets of plant health that are imperceptible to the unaided eye.
- **Improved Identification of Plant Stress:** Facilitates the early recognition of plant stress. This include challenges such as nutritional deficits, water scarcity, and disease. Stressed plants exhibit superior light reflectance compared to healthy plants. Multispectral cameras can identify these differences, providing early alerts of possible problems.
- **Vegetation Indices:** Agriculturalists utilize multispectral data to compute vegetation indices. An example is the Normalized Difference Vegetation Index (NDVI). This offers accurate insights about agricultural health and vitality. NDVI maps derived from Multi-Spectral Imaging delineate regions of elevated vegetation density. It also emphasizes regions of sparse vegetation. This enables farmers to make better informed decisions.
- **Data Analytics and Processing:** The DJI Mavic 3 Multi-Spectral Drone acquires data through analytics and processing. This dataset comprises multispectral and thermal photos. Sophisticated software analyzes this data. This type of software handles raw imagery. It produces comprehensive maps and analyses of crop vitality. This transformation transforms the data into useful insights. Sophisticated software possesses robust data analytics capabilities. These instruments provide thorough examination of agricultural conditions. This capacity enables farmers to make educated decisions. It enhances agricultural productivity. It also results in improved resource management.

Crop Monitoring Using Drones

Precision Agriculture

The utilization of drones for crop monitoring is equivalent to the principle of precision agriculture. Drones outfitted with sophisticated sensors and imaging technology furnish farmers with immediate data regarding crop health, soil conditions, and possible problems, enabling precise interventions and enhanced resource management.

Benefits of Crop Monitoring with Drones

The benefits of employing drones for agricultural surveillance are diverse. The capacity to obtain high-resolution images and data enables the early identification of crop stress, pathogens, and nutritional deficits. This early insight enables farmers to implement corrective actions swiftly, reducing potential losses and enhancing overall crop health.

Resource Optimization

Drones enhance resource efficiency by delivering farmers insights into the precise requirements of their crops. The data gathered by drones facilitates accurate and effective application of water, fertilizers, and pesticides, hence decreasing waste and mitigating the environmental impact linked to excessive input usage.

Monitoring the Overall Helth

Assessing the overall health and vitality of crops is crucial for optimizing yields and quality. Agricultural drones offer critical insights into crop health by acquiring multispectral imagery that can disclose differences in chlorophyll content, biomass, and stress levels. Through the analysis of this data, farmers may evaluate the status of their crops and adjust management measures accordingly.

Weed infestation is a considerable burden for farmers, impacting crop yields and profitability. Agricultural drones utilizing machine learning techniques and artificial intelligence can precisely identify and categorize weeds in agricultural fields. Equipped with this information, farmers can execute tailored weed management tactics, like precision herbicide application or mechanical cultivation, thereby diminishing dependence on chemical inputs.

- **Sensor Technology:** Drones are outfitted with an array of sensors that collect diverse data types. Commonly utilized sensors for monitoring plant health encompass multispectral and hyperspectral cameras, infrared cameras, LiDAR (Light Detection and Ranging), and GPS receivers.
- **Data Acquisition:** Drones traverse designated flight patterns over agricultural or wooded regions, collecting data during their flight. The onboard sensors gather data regarding foliage color, temperature, and height. Multispectral and hyperspectral cameras can catch reflected light from vegetation, offering insights into their health and stress conditions.
- **Data Analysis:** The data acquired by drones is subsequently examined with specialist software. This investigation offers insights into diverse plant health

indices, including chlorophyll content, moisture levels, pest and disease infestations, and overall crop vitality.

- **Proactive Identification:** A notable benefit of drone-assisted surveillance is proactive identification. Early detection of concerns such as pest infestations or water stress in crops enables farmers to implement corrective measures prior to substantial damage.

Figure: 7.3 Start-ups Specializing in Drone Technology are Encouraging Farmers to Use Drones for Monitoring Crops and Spraying Fertilizers and Pesticides

- **Precision Agriculture:** is facilitated by drone technology. Farmers can focus on particular sections of their fields requiring treatment, instead than distributing resources uniformly throughout the entire area. This diminishes the application of herbicides and fertilizers, yielding financial savings and ecological advantages.
- **Forest Management:** Drones are utilized in forest management to evaluate tree health, monitor for diseases or pests, and strategize timber harvests. Drones can generate intricate 3D models of forests, which are essential for resource management.
- **Research and Conservation:** Researchers and conservationists employ drone-based plant health monitoring for diverse purposes, including as examining the effects of climate change on plants, monitoring endangered species, and assessing ecosystem alterations over time.
- **Regulatory Considerations:** Drone operations, such as plant health monitoring, are governed by authorities in numerous nations. These restrictions address matters including flight altitude, privacy, and safety. Operators must recognize and adhere to local regulations.
- **Data Integration:** The data acquired by drones can be amalgamated with additional information sources, including meteorological data, satellite imaging, and terrestrial sensors, to furnish a more holistic perspective on plant health and environmental circumstances.

Drone-assisted plant health monitoring is an effective instrument that enhances agricultural efficiency, minimizes resource consumption, and promotes sustainable land management techniques. Nonetheless, it necessitates a specific degree of technological proficiency to operate the drones and analyze the gathered data efficiently. As technology progresses, the applications and functionalities of drone-based plant health monitoring are expected to broaden significantly.

APPLICATIONS OF DRONE-BASED CROP HEALTH MONITORING

- **Early Identification of Agricultural Issues:** Drones facilitate the prompt recognition of potential complications. This include infections, pest infestations, and nutritional deficiencies. Multi-Spectral Imaging, specifically, aids in detecting these problems prior to their proliferation. This enables farmers to implement proactive strategies to safeguard their crops.
- **Precision Agriculture:** Drone-based Crop Health Monitoring provides information that enhance precision agriculture methods. Agriculturalists may administer fertilizers, pesticides, and irrigation as required. This minimizes waste and enhances the efficiency of input utilization. Multispectral imaging facilitates precise evaluations of agricultural requirements. This facilitates the use of inputs according to these specific assessments.
- **Yield Estimation:** Drones facilitate the assessment of crop yields through the analysis of plant health. They also analyze growth patterns during the growing season. Multispectral imaging enhances the precision of yield forecasts. This enables farmers to optimize their harvest planning and resource management.
- **Optimized Agricultural Management:** Drone-assisted Crop Health Surveillance provides real-time information on field conditions. This facilitates ongoing modifications to agricultural management strategies. The integration of high-resolution photography and multispectral data facilitates dynamic decision-making. This results in enhanced agricultural productivity and profitability.

Yield Estimation

Agricultural drones assess agricultural yield by taking photographs and data regarding crops and their surroundings, subsequently evaluating this data to forecast the output.

- Drones, equipped with superior cameras, acquire exceptional images of fields, facilitating the creation of thorough and precise field maps that greatly enhance planning and management.
- These maps are utilized for multiple purposes, including determining field area, assessing crop health variations, and designing irrigation systems.
- Drones capture data from various angles while flying over fields, producing 3D topographical models of agricultural land.
- This model is essential since it provides extensive data for the design of effective irrigation and drainage systems, reducing water consumption and alleviating erosion problems.

Key Steps Involved in Drone-based Crop Yield Estimation

- **Flight Planning:** Strategically devise the drone's trajectory across the agricultural expanse, taking into account variables such as field dimensions, crop species, and the requisite data resolution.
- **Data Acquisition:** Equip the drone with suitable sensors, including multispectral or hyperspectral cameras, thermal sensors, or LiDAR, to obtain high-resolution photos, temperature data, and other pertinent information regarding the crops and the surrounding environment.
- **Processing:** Employ specialist tools to analyze the acquired data and generate comprehensive maps and models of the agricultural region. This may entail concatenating images, conducting image analysis, and amalgamating data from diverse sensors to provide comprehensive datasets.
- **Image Analysis and Interpretation:** : Evaluate the acquired imagery to assess crop vitality, identify regions of stress or disease, track growth trends, and detect any potential factors that may impact crop production.
- **Yield Estimation:** Utilize the gathered data and analytical outcomes to assess agricultural yield, considering variables such as plant density, biomass, and vegetation indices. Advanced algorithms and machine learning methodologies can be utilized to enhance yield prediction accuracy.
- **Decision Support:** Furnish farmers with pragmatic insights and suggestions derived from yield estimation data, empowering them to make informed decisions regarding irrigation, fertilization, pest management, and comprehensive crop management plans.
- **Integration with Farm Management Systems:** Incorporate the gathered data and insights into current farm management systems or software platforms to enhance decision-making processes and augment overall farm production.

Drone-assisted agricultural yield estimation has the capacity to transform contemporary agriculture by facilitating accurate and prompt crop monitoring, optimizing resource distribution, and improving overall crop management strategies. It can substantially enhance sustainable agricultural methods and assist in fulfilling the increasing worldwide need for food production.

Revolutionizing Crop Yield Prediction with Drones

Historically, crop production forecasting has predominantly depended on terrestrial surveys and past meteorological data. Ground surveys, although yielding localized data points, are labor-intensive, time-consuming, and constrained in scope. They may not reflect the comprehensive health and variability of an extensive field. Likewise, historical weather data provides merely a general indication of prospective yield and fails to consider real-time variables that might profoundly affect crops, such as localized disease outbreaks or abrupt fluctuations in moisture levels.

Drones integrated with GIS technology provide an innovative method for predicting crop yields. These unmanned aerial vehicles (UAVs) can traverse fields, acquiring high-resolution data that offers an extensive overview of crop health and growth. Principal technologies utilized in these drones comprise:

- **Multispectral and Hyperspectral Imaging:** These sensors acquire data beyond the visible spectrum, disclosing information regarding plant health that is imperceptible to the unaided eye. These sensors can identify small fluctuations indicating stress, nutrient shortages, or disease presence by monitoring the reflectance of various wavelengths of light.
- **LiDAR (Light Detection and Ranging):** LiDAR (Light Detection and Ranging) use laser pulses to generate intricate three-dimensional maps of the terrain. This facilitates accurate assessment of plant height, canopy coverage, and overall field topography. This data is essential for comprehending differences in solar exposure, water drainage, and possible microclimates within the field.

The data acquired by geospatial drones is extensive and intricate. Advanced data analysis techniques are utilized to get significant insights. Machine learning techniques are essential for processing pictures taken by drones and LiDAR data. These algorithms may discern patterns and correlations among the acquired data and variables such as plant health, soil conditions, and historical yield. Through the analysis of these correlations, one can:

- Determine particular elements that are influencing crop health and yield positively or adversely.
- Construct accurate yield forecasting models that use current field conditions and historical data.

Benefits of Drone-based Crop Yield Prediction

The amalgamation of drones and geospatial technologies provides numerous advantages for forecasting crop yields. These consist of:

- **Enhanced Precision and Turnaround Time:** In contrast to conventional techniques, drone-based forecasting provides much superior precision in yield predictions. The capacity to collect real-time, high-resolution data facilitates the prompt detection of possible problems, empowering farmers to implement corrective measures prior to substantial yield damage.
- **Targeted Resource Allocation:** Accurate production forecasts enable farmers to deploy resources, like fertilizers and water, with precision. This enhances resource efficiency and minimizes waste, hence augmenting farm profitability.
- **Minimized Crop Losses:** By facilitating early identification of crop stress or disease, drone-based forecasting enables farmers to implement prompt remedies. This can markedly diminish crop losses and provide a more uniform harvest.

Drones, outfitted with geospatial technologies, are radically transforming the domain of crop yield forecasting. This system offers an extensive overview of field conditions and utilizes advanced data processing capabilities, enabling farmers to enhance their methods and increase their yields.

THE FUTURE OF THE SYNERGY BETWEEN DRONES AND GEOSPATIAL IN AGRICULTURE

The future of geospatial technology and drones is replete with exhilarating prospects. Ongoing improvements in drone technology and geospatial data analysis methodologies are poised to enhance applications in agriculture, disaster management, and other fields.

Ongoing Advancements

- **Drone Technology:** Drones are evolving in sophistication, featuring extended flight durations, increased cargo capacity, and improved maneuverability. Furthermore, improvements in battery technology are prolonging flight durations, enabling the acquisition of more extensive data sets. Autonomous drone operations are being investigated to enhance data collection efficiency.
- **Geospatial Data Analysis:** The domain of geospatial data analysis is perpetually advancing. Machine learning algorithms are increasingly sophisticated, enabling more profound insights to be derived from data gathered by drones. Moreover, the incorporation of artificial intelligence (AI) is expected to enhance the automation of data processing operations and produce real-time actionable insights.

Potential Future Applications

- **Precision Agriculture Utilizing Real-Time Data:** The real-time data acquired by drones might be employed for exceptionally accurate agricultural methodologies. Through the ongoing surveillance of crop vitality and environmental parameters, farmers can modify irrigation, fertilization, and pest management tactics on a field-specific, and even plant-specific, level. This degree of precision can enhance resource usage and optimize agricultural output.
- **Early Warning Systems for Natural Disasters:** Drones outfitted with sophisticated sensors can be utilized for proactive control of natural disasters through Early Warning Systems. They can be utilized to surveil regions susceptible to landslides, floods, or wildfires, facilitating the early identification of possible hazards. This real-time data can be incorporated into current early warning systems, allowing authorities to issue prompt evacuation orders and commence mitigating activities.
- **Integration with Other Geospatial Platforms:** The data acquired by drones may be effortlessly integrated with many geospatial platforms, including Geographic Information Systems (GIS). This will facilitate the development of intricate, multi-dimensional visualizations that integrate drone footage with pertinent geographic data (e.g., historical meteorological patterns, soil composition maps). This extensive perspective might enable stakeholders across diverse domains to make informed decisions grounded in a deeper comprehension of the geographical context.

The extensive utilization of drone technology requires meticulous attention to ethical dilemmas and data protection issues. Regulations must be instituted to guarantee responsible drone operation, addressing issues such as privacy infringements and the potential exploitation of data gathered by drones. Engaging in open discourse and collaborative initiatives can facilitate the responsible utilization of this technology's advantages.

Disease Detection

Drone technology utilizes an automated process that commences with the acquisition of photos of infected plants through diverse sensors and cameras. Subsequent to feature extraction, image processing methodologies employ suitable conventional machine learning or deep learning algorithms.

The integration of AI drones in agriculture will yield significant advantages for the agricultural sector. Accurate identification of crop diseases by drone photography and AI algorithms facilitates prompt treatments, curtailing further dissemination and reducing crop losses. This results in enhanced yields, elevated incomes, and greater economic stability for farmers. Furthermore, minimizing chemical application through targeted interventions fosters sustainable agricultural practices, safeguarding environmental integrity for future generations.

Their initiative centers on intelligent crop disease identification with AI drones. Drones utilize artificial intelligence and computer vision to detect, monitor, and analyze agricultural illnesses. The deep learning system incorporated into the drones facilitates accurate detection of illness patterns, while the computer vision algorithms identify the specific locations of the damaged areas. Additionally, the drones offer fertilizer suggestions customized to the particular ailment, equipping farmers with practical information for efficient disease management.

INTEGRATION OF DRONE DATA WITH PRECISION AGRICULTURE TECHNIQUES

In most drought-prone regions, supplying adequate water for crop cultivation is a formidable challenge. Currently, existing irrigation methods deliver water to a field uniformly. It is essential to supply water precisely where it is needed to ensure the rational and optimal utilization of limited water resources. Water is a vital input for crop growth; insufficient or excessive water can hinder normal plant development. Agriculturalists must implement effective irrigation management tailored to the specific crops farmed in extensive farmland. The drone aids farmers in making informed judgments for precise irrigation management with thermal digital cameras. The drone can assess soil moisture stress and excess water in specific agricultural areas. The field survey conducted with the drone enhances water use efficiency and facilitates the discovery of irrigation leaks through regular monitoring.

Drones outfitted with sensors and thermal and multispectral imaging cameras capable of detecting heat and water stress in crops within a specific agricultural area. Irrigation can be administered to crops according to their requirements. This would prevent water wastage and guarantee the efficient utilization of irrigation water. Drones were employed in Colombo by the International Water Management Institute (IWMI) to monitor rice cultivation in the water-deficient Anuradhapura region. The institute captured images of the paddy fields utilizing near-infrared and red, green, and blue sensors. In Africa's rain-fed agricultural system, water is the critical determinant. UAV technology is facilitating the planning and building of a viable irrigation system.

- In irrigation management, water is a vital resource in agriculture for promoting healthy crops and maximizing yields. Drones are utilized to monitor soil moisture levels across fields with high precision.
- Drones in agriculture are extensively utilized due of their multispectral sensor capabilities. This facilitates the collection of data regarding soil moisture levels and plant water stress.
- The acquired data is utilized to create maps indicating regions of elevated or diminished moisture levels, facilitating farmers in employing precision irrigation methods like drip irrigation, hence conserving water and reducing expenses.
- This focused strategy guarantees effective water utilization, minimizing waste and enhancing crop development. Effective water management is crucial for sustainable agriculture, particularly in regions susceptible to drought.

Precision Spraying

We recognize that conventional techniques for applying insecticides, herbicides, and fertilizers frequently entail indiscriminate spraying over large areas.

- Drones, now an invaluable technology in agriculture, improve the administration of fertilizers, insecticides, and herbicides through precise accuracy.
- Drones are equipped with specialized sprayers and sensors that apply pesticides, herbicides, and fertilizers directly to specific locations.
- This mitigates waste and diminishes the likelihood of overspray affecting non-target zones or aquatic environments. Additionally, it diminishes farmers' exposure to hazardous pesticides and enhances the general health of crops.

Figure: 7.4 Precision Application of Spray

Pests, weeds, and plant diseases result in crop losses of approximately 15-25% of total food production in India. According to the Food and Agriculture

Organization (FAO), if farmers discontinue the use of crop protection technologies, they may incur yield losses of around 40% due to pests and plant diseases. The conventional method of applying chemicals in agriculture is highly labor-intensive and presents health hazards to workers due to exposure to toxic substances affecting the eyes, lungs, and skin. Approximately 300,000 farmers experience respiratory ailments as a result of pesticide application. Various studies indicate that pesticide usage resulted in the deaths of 183 farmers in Maharashtra from 2013-14 to 2017-18, and 442 nationwide. Additionally, snakebites are a significant cause of mortality among farmers in India. Approximately 58,000 farmers lose their lives. Drone technology has the ability to mitigate these issues in agriculture.

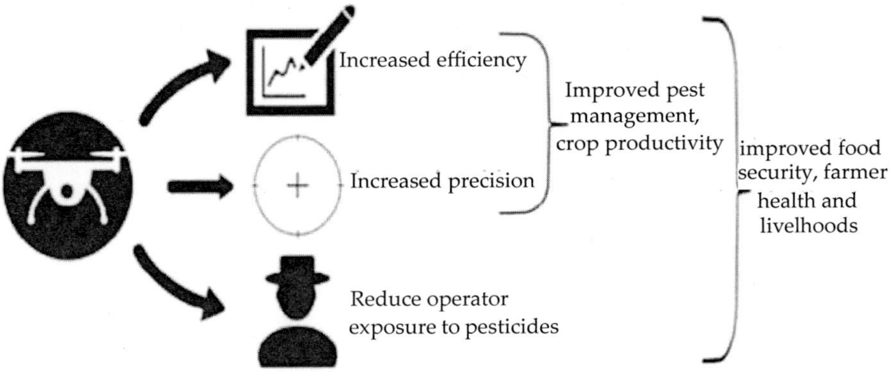

Figure: 7.5 Benefits of drone for pesticide application

Ensuring the application of chemicals in a specified dosage is challenging and poses risks in tough geographical regions, such as steep areas with slopes. Drone technology can address these challenges by integrating AI, machine learning, and the Normalised Difference Vegetation Index (NDVI) to gain accurate insights into soil conditions, plant health, and yield predictions. High-quality cameras and sensors affixed to the drone facilitate site-specific spraying. The drone is capable of scanning the agricultural area for spraying and assures the application of a certain quantity of liquid fertilizers and pesticides on the scanned crops. Research on drone spraying indicates that drones can operate five times more rapidly than conventional agricultural spraying methods, thereby conserving time and lowering input costs. Research indicates that map-based optimized nitrogen application leads to a 20% decrease in nitrogen intake.

The anticipated decrease in agrochemical input costs (pesticides/weedicides) is around 25-30% as a result of automation. For example, the application of nano urea, priced at Rs.240 per bottle for spraying on an acre, results in a 25% savings, hence contributing to an overall reduction in input costs. Moreover, drones can reduce agricultural water consumption by up to 90% due to their minute droplet sizes of about. 50 microns in comparison to the approximate size of hand spray droplets. 500 micrometers. Drone spraying conserves time, requiring approximately 5-7 minutes per acre, while conventional spraying necessitates a whole day to cover 3-4 acres.

Planting and Seeding

Automated drone systems provide exceptional precision in seed planting. These drones guarantee consistent seed dispersal, so minimizing waste and fostering uniform crop development.

- This method facilitates cultivation in regions that would often be challenging or unfeasible to farm.
- Drones has the capability to cultivate more accurately and efficiently on inclines or inaccessible regions that pose difficulties for conventional machinery.
- Agriculturists formerly employed heavy machinery to compact the earth; however, they now utilize drones that apply a gentler approach, reducing soil compaction and fostering robust root growth.

DECISION SUPPORT SYSTEMS FOR CROP MANAGEMENT BASED ON DRONE DATA

Decision support systems (DSS) for crop management with drone data are advanced instruments that assist farmers in making informed, data-driven decisions to enhance agricultural operations. Drones, outfitted with diverse sensors such as multispectral, hyperspectral, and thermal cameras, are employed to gather real-time data regarding the status of crops, soil, and environmental variables. These systems evaluate the data to deliver insights that facilitate informed decision-making.

This is an overview of the functioning of Decision Support Systems (DSS) for crop management utilizing drone data.

Data Collection with Drones

- **Multispectral Imaging:** Drones utilize multispectral imaging to acquire photos that assess light reflected from vegetation across many wavelengths. These images facilitate the evaluation of crop health by detecting early indicators of stress, nutrient shortages, or disease.
- **Thermal Imaging:** Drones outfitted with thermal cameras identify temperature fluctuations in crops, signaling water stress or regions requiring irrigation.
- **LiDAR and RGB Cameras:** These sensors are capable of generating three-dimensional maps of agricultural areas, evaluating plant height, identifying pest damage, and estimating biomass.
- **GPS and Georeferencing:** Drones utilize GPS for georeferencing acquired data, facilitating the tracking of specific field regions and the monitoring of temporal changes.

Data Processing

- **Preprocessing:** Raw drone data (images and videos) undergoes processing to eliminate noise and rectify distortions induced by weather or lighting conditions.
- **Image Analysis and Feature Extraction:** Specialized software evaluates images, extracting pertinent features such as the Normalized Difference

Vegetation Index (NDVI) for assessing plant health, and the Soil-Adjusted Vegetation Index (SAVI) for measuring the greenness and density of the crop canopy.

- **Data Integration:** Additional data sources, such as meteorological data and soil moisture levels, are amalgamated to present a comprehensive overview of field conditions.

Decision Support System (DSS) Algorithms

- **Crop Health Monitoring:** The technology employs machine learning and AI algorithms to evaluate crop health and forecast future growth patterns. This assists farmers in pinpointing regions requiring intervention, including those affected by pests, illnesses, or nutrient deficits.
- **Irrigation Management:** Utilizing temperature and soil moisture data, the Decision Support System (DSS) may advise on the optimal locations and quantities for irrigation, thereby conserving water and minimizing expenses.
- **Pest and Disease Detection:** The technology identifies anomalies or trends indicative of pest or disease presence, enabling farmers to implement timely interventions, such as targeted pesticide treatments.
- **Yield Prediction:** The system can forecast the anticipated yield of a crop by analyzing present conditions, historical data, and growth tendencies. This assists farmers in making decisions on harvesting, selling, and future crop planning.
- **Fertilization and Nutrient Management:** Through the analysis of soil and crop data, the system can prescribe optimal fertilizer application rates, enhancing crop quality and minimizing fertilizer overuse.

Actionable Recommendations

- **Field Mapping and Zone Management:** Utilizing drone footage, DSS may generate field zone maps that delineate areas with varying requirements (e.g., irrigation zones, nutrient deficiency zones, or insect hotspots). Farmers can subsequently allocate resources effectively by concentrating on particular locations, thereby minimizing waste.
- **Prescriptive Analytics:** The Decision Support System (DSS) provides explicit advice for farmers regarding the optimal timing for pesticide application, the requisite quantity of fertilizer, and the most effective irrigation schedules.
- **Real-Time Alerts:** The system is capable of dispatching notifications upon the detection of concerns (such as plant stress, pests, or illnesses), enabling farmers to respond promptly.

Benefits of Drone-Based DSS for Crop Management

- **Precision Agriculture:** Facilitates more precise, data-informed decisions, hence minimizing expenses and enhancing yields.
- **Decreased Input Expenditures:** Through improved control of irrigation, fertilizer, and pesticide application, farmers can minimize superfluous input expenses.

- **Enhanced Crop Yield and Quality:** Early identification and resolution of issues can substantially enhance crop yields and quality.
- **Sustainability:** Enhanced resource efficiency, including water and fertilizers, mitigates the environmental impact of agriculture and promotes sustainable farming practices.
- **Time Efficiency:** Real-time data enables farmers to implement immediate actions, hence enhancing overall farm productivity.

Examples of DSS Platforms Using Drone Data

- **Sentera:** A firm providing drone-enabled crop health assessment with real-time data for precision agriculture.
- **AgEagle:** offers solutions utilizing drone data for the monitoring of crop health, incorporating integrated machine learning algorithms to evaluate plant conditions.
- **CropX:** This technology integrates drone data with soil moisture sensors and meteorological information to enhance irrigation and fertilizing efficiency.
- **Skycatch:** offers drone-derived data analytics for the construction and agricultural sectors, encompassing crop health assessments and yield forecasting models.

Challenges and Future Outlook

- **Data Overload:** Analyzing substantial volumes of drone data can be computationally demanding, necessitating sophisticated tools and cloud-based systems.
- **Financial Implications and Accessibility:** The expense associated with drone technology and the requisite data analysis tools may restrict access for small-scale farmers.
- **Regulatory Constraints:** Certain areas impose limitations on drone operations, potentially restricting their application in agriculture.
- **Integration with Other Technologies:** DSS platforms must interact seamlessly with various agricultural technologies (e.g., IoT devices, meteorological stations), which can be intricate.

The prospects for Decision Support Systems (DSS) in crop management with drone data are promising, bolstered by breakthroughs in artificial intelligence, machine learning, and autonomous drones. As technology becomes increasingly affordable and accessible, the incorporation of these systems into routine agricultural processes is expected to expand, resulting in more sustainable and efficient farming methods globally.

Drone Imagery Analysis

Step 1: Photogrammetry Processing

Photogrammetry processing is utilized to convert unprocessed drone data into meaningful insights. This method employs drone footage to produce intricate

3D landscape models, crop surface maps, and volumetric biomass models. These outputs provide accurate geometric information regarding the crops and the land's topography.

Step 2: Machine Learning Classification

Machine learning classifiers, including convolutional neural networks, are essential to this process. Cloud-based platforms facilitate the expeditious processing of picture pipelines, enabling the effective management of hundreds of high-resolution photographs obtained during each drone flight. This phase is crucial for prompt and effective data processing.

Step 3: Geographic Information Systems (GIS) integration

By superimposing drone-collected data onto geographical maps, farmers and agricultural specialists can acquire a thorough comprehension of the terrain. This comprehensive strategy enables the implementation of effective measures derived from drone data analytics insights.

The Future of Digitized Agriculture with Drones

The incorporation of drones in agricultural activities is profoundly transforming the paradigms of farming, irrigation, and resource management. The real-time data collection capabilities of drones enhance precision agriculture by enabling data-driven decision-making instead of reliance on conjecture. These transformations are enhancing agricultural practices by facilitating improved monitoring of crop health and irrigation, yield forecasting, and soil management, among other aspects, making them more efficient and sustainable.

Considering the ongoing advancements in drone technologies, the future of agriculture appears increasingly promising. Utilizing these technology will enable agriculturalists to enhance their productivity, reduce costs, and contribute to environmentally sustainable agricultural practices. Consequently, improving agricultural practices through an intelligent farming methodology.

8

Using Drones for Early Pest and Disease Detection

INTRODUCTION TO DRONES USED FOR EARLY PEST AND DISEASE DETECTION

The utilization of drones for the early detection of pests and diseases in agriculture represents a novel and effective strategy that enables farmers to proactively oversee crop health. Drones outfitted with sophisticated sensors can consistently surveil fields, delivering real-time data and facilitating the prompt detection of possible dangers.

Agricultural drones are employed to identify pests and illnesses at an early stage by capturing aerial imagery of crops. This enables farmers to respond swiftly to safeguard their crops, perhaps resulting in increased yields and reduced environmental impact.

The utilization of drones in agricultural pest management presents numerous benefits compared to conventional techniques. Drones outfitted with infrared cameras can furnish farmers with real-time information on crop health, facilitating the early identification of possible problems such as pests, illnesses, or nutritional deficits.

Pathology can adversely affect plant vitality prior to the manifestation of any obvious indicators such as foliar discoloration. Although these pressures are imperceptible to the unaided eye, cameras equipped with specialized filters may identify these nuanced variations.

Researchers at the Department of Life Sciences and Computing are collaborating with the agricultural services firm Agrii to develop cameras placed on drones. These cameras can autonomously identify early signs of disease and notify farmers when to apply treatments, prior to crop harm.

Applying fungicides promptly in the appropriate locations would enable farmers to utilize them effectively and optimize their application. They are presently creating a method to predict Septoria, a significant fungus impacting wheat, but anticipate that the principle might subsequently be modified for other illnesses.

In recent years, drone technology has become a revolutionary instrument in agriculture, providing innovative methods for crop protection. Drones, or autonomous aerial vehicles, offer numerous advantages such as enhanced efficiency, diminished labor demands, decreased pesticide loss to the environment, and extensive area coverage. In recent years, the utilization of drones across many industries has increased significantly, with agriculture being no exception.

Drones have demonstrated adaptability and versatility as technologies capable of transforming crop protection methods. Integrating drones into crop protection methodologies enables farmers and agricultural experts to acquire critical insights and enhance efficiency in field monitoring and treatment. Sprayer drones have become a transformative force in agricultural techniques. These drones are pertinent not just to agricultural practices but also provide innovative and commercial methods for crop protection. Drones may readily reach the peripheries and muddy regions of rice paddies that are difficult to access on foot without harming the rice crop. Furthermore, the utilization of drones in agricultural protection facilitates more exact and precise application, mitigating the dangers of inconsistent application rates and variable walking speeds inherent in manual spraying.

The utilization of drones in agricultural pest management presents numerous benefits compared to conventional techniques.

- Drones outfitted with infrared cameras can furnish farmers with real-time information regarding crop health, facilitating the early identification of possible problems such as pests, illnesses, or nutrient deficits. This proactive strategy allows farmers to implement precise and timely measures to safeguard their crops, resulting in increased yields and less environmental impact.
- The utilization of drones in agricultural pest management might markedly decrease the quantity of pesticides and water required for field treatment. By accurately identifying treatment-required regions, farmers can reduce chemical usage while efficiently protecting their crops. This results in both cost savings and the promotion of sustainable agriculture methods.
- The data gathered by drones can be utilized to generate intricate maps of agricultural lands, offering farmers critical information into crop growth trends and the efficacy of various treatments. This knowledge can guide future decision-making and enhance resource allocation, efficiency, and conservation of energy and time.

In conclusion, the anticipated benefits are as follows:

- **Improved Efficiency and Precision:** Drones provide swift and accurate data acquisition across extensive agricultural regions, exceeding the effectiveness of manual scouting and satellite photography. This efficiency improves decision-making processes and facilitates prompt reactions to new hazards, hence enhancing crop yields and profitability.

- **Cost-Effectiveness:** Although initial investments in drone technology may appear considerable, the long-term advantages surpass the expenses for several farms. Drones provide a significant return on investment and enhance the economic sustainability of agricultural enterprises by minimizing labor demands, maximizing resource utilization, and mitigating yield losses.
- **Environmental Sustainability:** By reducing chemical inputs and implementing focused management strategies, drone technology promotes eco-friendly agriculture. Decreased pesticide application diminishes the potential for ecological damage and fosters biodiversity preservation, in accordance with sustainable agricultural practices and regulatory standards.

FACTORS AFFECTING THE EFFICIENCY OF DRONE USAGE FOR PLANT PROTECTION

The efficiency of a drone is contingent upon elements including droplet size, wind speed, flight speed, and altitude.

- Meteorological factors, including wind speed, temperature, and humidity, might influence the efficacy of pesticide applications under actual field circumstances.
- Lv et al. (2019) examined the efficacy of drones in a controlled environment to minimize external variables and assess the impact of drone flight speed on droplet properties, including size, shape, dispersion, and uniformity. The distribution of droplets and their size are crucial criteria for effective input delivery.
- **Droplet Dispersion:** An essential factor to evaluate for the effectiveness of drone-assisted spraying is the distribution of droplets. The criteria employed to assess droplet dispersion effectiveness include density, surface coverage, arithmetic mean droplet size, and coefficient of variation.
    - Droplet density denotes the number of droplets deposited per unit area, typically evaluated using blotting paper.
    - Surface coverage quantifies the aggregate area occupied by all deposited droplet particles per unit area.
    - The arithmetic mean droplet size denotes the average diameter of all droplets in a singular spray sample.
    - The coefficient of variation (CV) reflects the consistency of droplet distribution in an aerial spraying operation. A lower coefficient of variation (CV) indicates a more uniform distribution, which ensures effective coverage of the target area and improves pesticide efficiency.
- **Droplet drift:** When aerial spraying is performed using drones, a portion of the dosage may not reach the designated area, referred to as droplet drift or spray drift. To address this drift and reduce pesticide waste, a defined operational strategy for drone-assisted pesticide application includes appropriate meteorological conditions. Research indicates that an optimal flight altitude of 1.5–2.0 meters and a speed of 5 meters per second, along with a wind velocity under 5 km/h, facilitate effective pesticide application

using drones. Moreover, research examining the correlation between the coefficient of variation (CV) of droplet distribution and UAV flight altitudes in cotton fields has demonstrated that a flight height of 2 meters produces more favorable outcomes than a height of 1.5 meters.

Applications of Drone Technology in Crop Protection

- **Aerial Surveillance:** Drones outfitted with high-resolution cameras and multispectral sensors allow farmers to do aerial assessments of their fields with unparalleled precision and efficiency. These surveys enable the prompt identification of pests, illnesses, nutrient deficits, and other stresses, facilitating timely intervention.
- **Accurate Application of Inputs:** By incorporating drones with precision agricultural technologies, farmers can accurately administer pesticides, herbicides, fertilizers, and other inputs based on real-time field data. This focused strategy diminishes chemical application, curtails environmental contamination, and enhances resource efficiency.
- **Pest Detection and Management:** Drones outfitted with advanced sensors, like LIDAR and infrared cameras, may identify pests and their damage with greater precision than conventional scouting techniques. This early detection facilitates proactive pest control tactics, encompassing localized actions and targeted pesticide applications, so minimizing crop losses and decreasing dependence on broad-spectrum pesticides.

Challenges and Limitations

- **Technical Complexity:** The operation of drones and the analysis of the extensive data they produce necessitate particular skills and experience. A significant number of farmers may be deficient in the technical expertise or resources necessary to properly employ drone technology, underscoring the necessity for training programs and accessible solutions.
- **Data Integration and Interpretation:** The incorporation of drone-acquired data into established farm management systems poses difficulties, especially for small-scale farmers with restricted access to digital resources. Furthermore, the interpretation of intricate remote sensing data necessitates sophisticated analytical skills and may present obstacles to adoption for certain users.
- **Regulatory Challenges:** Despite the potential advantages, the extensive implementation of drone technology in agriculture encounters regulatory obstacles concerning airspace rules, privacy issues, and certification mandates. Explicit standards and efficient approval procedures are essential for the incorporation of drones into agricultural practices.

TYPES OF SENSORS FOR EARLY DETECTION OF PEST IN CAEPS

Drones are outfitted with an array of sensors capable of capturing several data kinds, essential for identifying pests and illnesses in crops.

- **Multispectral Sensors:** These devices catch light reflected from crops across several wavelengths, including those beyond the visible spectrum. Analyzing the reflection of light by plants in the infrared or near-infrared spectrum enables the early detection of stress in crops, maybe induced by pests or illnesses. Indices such as the Normalized Difference Vegetation Index (NDVI) and the Normalized Difference Red Edge (NDRE) facilitate the assessment of plant health.

- **Thermal Cameras:** Thermal imaging detects temperature fluctuations in crops. Plants experiencing stress due to pests, illnesses, or insufficient water will generally exhibit distinct temperature profiles in contrast to healthy plants. This can identify sections of the field that may require attention.
- **Hyperspectral Imaging:** These sensors acquire an extensive spectrum of wavelengths, significantly broader than that of multispectral sensors, yielding comprehensive data regarding plant health. Hyperspectral data can identify subtle changes in the spectral signature of plants, perhaps signaling early indications of disease or pest damage.
- **RGB Cameras:** Standard RGB cameras on drones can collect visible images that, when integrated with image processing, provide the detection of alterations in leaf color or texture, frequently indicating the initial signs of pests or illnesses.

How Drones Detect Pests and Diseases

Drones capture large volumes of data from the field, which is then processed and analyzed to identify early signs of pest infestations or diseases.

- **Change Detection:** Drones can identify alterations in crop appearance through continuous aerial surveillance. Pests frequently inflict damage that leads to discoloration, inhibited development, or wilting. Disease outbreaks may similarly result in discolouration, lesions, or irregular growth patterns.
- **Pattern Recognition:** Machine learning and AI algorithms evaluate data acquired by drones to discover spatial patterns characteristic of specific pests or diseases. A consistent patch of yellowing leaves may signify a fungal infection or nutrient shortage resulting from insect activity.
- **Hotspot Identification:** Drones can discover "hotspots," or regions with elevated pest populations or visible disease symptoms. These hotspots can be prioritized for treatment, thereby diminishing overall pesticide application and enhancing efficiency.

Benefits of Drone-Based Pest and Disease Detection

- **Early Detection:** Prompt identification of pests or diseases facilitates prompt intervention, averting larger outbreaks and diminishing the necessity for costly treatments. This can avert substantial crop damage or productivity reduction.
- **Accuracy and Targeting:** Drones facilitate the generation of intricate maps of contaminated regions. This enables farmers to administer insecticides or fungicides solely to the impacted areas, hence decreasing pesticide usage and mitigating environmental consequences.

- **Minimized Chemical Application:** Prompt identification allows for more precise treatments, hence decreasing chemical usage. This is advantageous not only for the environment but also for diminishing the expenses associated with inputs such as insecticides.
- **Cost Efficiency:** Early detection of pests and diseases enables farmers to avert extensive damage, hence reducing the substantial expenses linked to pest management and crop loss.
- **Non-Invasive Monitoring:** Drones offer a non-invasive method for monitoring extensive agricultural areas, preserving the integrity of the environment and the crops, so promoting sustainability.
- **Consistent and Systematic Surveillance:** Drones can traverse agricultural fields routinely, delivering an ongoing influx of data. This enables farmers to regularly assess the condition of their crops, ensuring they do not overlook minor changes that may indicate a developing issue.

DRONE DATA PROCESSING FOR PEST AND DISEASE DETECTION

Subsequent to data collection by drones, the photos and sensor information require processing and analysis. This is the conventional method employed:

- **Image Processing Software:** This specialized software analyzes images collected by drones to identify areas of concern, such discolored spots in crops or alterations in vegetation health.
- **Machine Learning Algorithms:** AI models can be trained to identify patterns linked to pests and diseases. These models evaluate extensive information and can differentiate between healthy and stressed plants by their spectral signatures.
- **Early Warning Systems:** Certain sophisticated systems provide real-time notifications to farmers when algorithms identify anomalies in crop data, including indications of insect activity or disease outbreaks.

Examples of Applications in Pest and Disease Detection

- **Crop Health Surveillance for Fungal Pathologies:** Drones can identify early indicators of fungal infections such as powdery mildew or rust. Fungal diseases generally induce alterations in leaf texture and pigmentation, which drones can detect and assess.
- **Insect Infestation Surveillance:** Drones can detect variations in plant health or utilize infrared sensors to pinpoint regions with elevated pest activity. The presence of aphids or caterpillars can induce minor alterations in crop temperature or chlorophyll levels detectable by drones.
- **Weed Detection:** Drones outfitted with multispectral sensors facilitate the identification of weed populations in crops, which frequently vie for nutrients and water, potentially intensifying insect and disease issues.

Challenges and Considerations

- **Data Overload:** The substantial quantity of data amassed by drones necessitates considerable processing capabilities, posing a barrier for small farms lacking access to cloud computing or sophisticated software.
- **Weather Conditions:** Drones are affected by weather conditions, including wind, rain, or fog, which may impede data collection. Drones must operate at the proper altitude and angle to gather precise data, and weather conditions can disrupt this process.
- **Training and Expertise:** Proficient operation of drones for pest and disease identification necessitates specialized training to successfully manage the drones and accurately analyze the gathered data. Agriculturalists must comprehend how to interpret and respond to data to achieve efficacy.
- **Regulatory Restrictions:** In certain areas, restrictions may inhibit drone operations, potentially restricting the frequency and scope of drone surveillance.

Future Directions

- **Integration with Other technology:** Merging drone data with additional precision agricultural technology, including IoT sensors and satellite photography, can yield a more holistic perspective on crop health.
- **Autonomous Pest Control:** In the future, drones may possess the capability to identify pests and diseases as well as administer localized treatments autonomously, thereby augmenting the effectiveness of pest management.
- **Artificial Intelligence and Deep Learning:** With progress in artificial intelligence and machine learning, drone systems will enhance their accuracy in identifying early indicators of pests and diseases, potentially enabling real-time decision-making.

IDENTIFICATION OF COMMON PESTS AND DISEASES IN CROPS

"Diseases diminish yields at a time when high production is essential due to the increasing global population requiring sustenance. Mitigating yield loss from diseases such as Septoria will enable more efficient food production on reduced land, benefiting the agricultural sector, the populace, and the environment."

The drones will employ multispectral cameras that utilize specialized filters to catch reflected light from certain parts of the electromagnetic spectrum. Utilizing multiple lenses and various filters concurrently to capture reflected light enables scientists to examine the differential reflection of objects across the electromagnetic spectrum. Stressed plants generally have a 'spectral signature' that differentiates them from healthy ones.

Drones are exceptionally efficient in identifying pests and diseases. They utilize real-time data and artificial intelligence to promptly discover concerns. This facilitates prompt interventions.

AI drones can modulate spraying quantities according to crop requirements, enhancing insect management. AI-driven weed management systems have lowered pesticide application by as much as 90%. This not only reduces chemical usage but also benefits the environment.

Application	Outcome	Accuracy/Success Rate
AI Disease Detection	Apple Scab Presence	95%
AI Disease Detection	Yellow Rust in Wheat	High Accuracy
Automated Weed Control	Reduction in Herbicide Usage	Up to 90%
Precision Mapping	Detailed Terrain Analysis	1-5 cm/pixel resolution

Table: Important Pests and Diseases, their Symptoms and Control Measures

Pest/disease	Pest	Symptoms	Control measures
Insects	Thrips (Thrips tabaci)	1. Thrips infestation at an early stage (transplanting to 45 days) can be identified by curling and twisting of leaves. 2. Typical symptom is the presence of white or silvery patches on the leaves. 3. In severe infestation, whole plant looks blemished and turns white.	1. Planting of two rows of maize or one outer row of maize and an inner row of wheat as a barrier crop surrounding garlic crop (250 sq. m) at least 30 days prior to planting of garlic helps to block the movement of adult thrips. 2. Spray insecticides Profenofos (0.1%), Carbosulfan (0.2%) or Fipronil (0.1%) when thrips population crosses the economic threshold level of 30 thrips/plant.
	Eriophyid mite (Aceria tulipae)	1. Leaves do not open completely. Whole plant shows stunting, twisting, curling and yellow mottling. 2. Mottling is seen mostly on the edges of the leaves.	Spray Dicofol (0.2%) or sulphur (0.05%) as soon as the symptoms appear. Repeat the spray after 15 days, if necessary.

Pest/disease	Pest	Symptoms	Control measures
	Red spider mite (Tetranychus urticae)	1. White discoloration of leaves with small patches of faeces, eggs, larvae and adult mites. 2. Yellow or bronze spots on the leaves. 3. Spider webbing on leaves	Spray Dicofol (0.2%) or sulphur (0.05%) as soon as the symptoms appear. Repeat the spray after 15 days, if necessary.
Nematode	Stem and bulb nematode (Ditylenchus dipsaci)	1. D. dipsaci enters through stoma or plant wounds and create galls or malformations in plant growth. This allows for the entrance of secondary pathogens like fungi and bacteria. 2. Symptoms are stunted growth, discoloration of bulbs, and swollen stems.	1. Bulbs that show signs of disease should not be planted 2. Proper sanitation of fields and tools is essential because this nematode can survive and reproduce in infected plants and residues.
Fungal diseases	Purple bloch (Alternaria porri)	1. Initially small, elliptical lesions or spots on leaves that often turn purplish-brown and are surrounded by chlorotic margin. 2. If the spots enlarge, chlorotic margin extend above and below the actual lesion. Lesions usually girdle leaves, causing them to fall over. Lesions may also start at the tips of older leaves.	Spay Mancozeb (0.25%) or Tricyclazole (0.1%) or Hexaconazole (0.1%) or Propiconazole (0.1%) at 10-15 days intervals from 30 days after planting or as soon as disease appears
	Stemphylium blight (Stemphylium vesicarium)	1. Small yellow to orange flecks or streaks develop in the middle of the leaf which soon develop into elongated, spindle shape to ovate elongate diffused spots surrounded by characteristic pinkish margin.	Spay Mancozeb (0.25%) or Tricyclazole (0.1%) or Hexaconazole (0.1%) or Propiconazole (0.1%) at 10-15 days interval from 30 days after planting or as soon as disease appears

Pest/disease	Pest	Symptoms	Control measures
		2. The spots progress from the tip to the base of the leaves. The spots coalesce into extended patches, blighting the leaves and gradually the entire foliage.	
	White rot (Sclerotium cepivorum)	1. Yellowing and dying back of leaf tips. 2. Roots are normally destroyed. Superficial white fluffy mycelium may be present on decaying scales. 3. Brown or black sclerotia develop on surface or within tissue.	1. Crop rotation should be followed, infected plant may be destroyed and the soil around these plants should be treated. 2. Soil solarization at high temperature reduces the disease incidence. 3. Carbendazim (0.1%) may be applied.
Viral diseases	Onion yellow dwarf virus (OYDV)	Mild chlorotic stripes to bright yellow stripes, mosaic, curling of leaves and stunted growth	1. Use virus-free planting material 2. Foliar spray of insecticides Profenofos (0.1%) or Carbosulfan (0.2%) or Fipronil (0.1%) for controlling aphids, the virus vector
	Leek yellow stripe virus (LYSV)	1. Causes light yellow striping on the distal part the leaves, which can lead to dwarfing of the entire plant 2. The virus also causes bulbs to be smaller and malformed, which results in yield loss and post harvest storage loss	1. Use virus-free planting material 2. Foliar spray of insecticides Profenofos (0.1%) or Carbosulfan (0.2%) or Fipronil (0.1%) for controlling aphids, the virus vector.
	Irish Yellow Spot Virus (IYSV)	1. Straw-coloured, spindle-shaped spots with poorly defined ends on the leaves 2. These spots coalesce to form larger patches on the leaves 3. Clearly visible on older leaves	1. Eliminate volunteers, culls, and weeds in and around garlic fields. 2. Use virus-free planting material

Pest/disease	Pest	Symptoms	Control measures
			3. Avoid crop stress. 4. Thrips control may provide some reduction in Iris yellow spots as thrips are vectors of virus
Storage diseases	Penicillium rot (Pencilliumc orymbiferum)	1. Affected bulbs are light in weight and individual cloves are soft, spongy and powdery dry. 2. In an advanced stage of decay, the cloves break down in a green or gray powdery mass.	Store under dry conditions in storage structures to reduce rot development
	Black mould (Aspergillus spp.)	1. Bulbs show black discoloration at the neck and streaks of black mycelium and conidia beneath the outer dry scales. 2. In advanced stages all scales are infected and the bulb shrivels.	1. After proper drying, store bulbs in dry conditions. 2. Avoid bruising when bulbs are harvested, stored or transported.
	Fusarium basal rot (Fusarium oxysporum cepae)	1. Shattering of the cloves 2. Bulbs appear spongy or sunken and / or brown and watery when cut open. 3. White or light pink growth appear over the cloves	1. Do not store damaged bulbs. 2. Store bulbs in dry conditions with good ventilation 3. Follow crop rotation with non-host crops such as wheat and cowpea for 3-4 years
	Bacterial rot (Erwinia spp. and Pseudomonas spp.)	Watery, foul-smelling fluid squeezes from the neck of diseased bulbs when pressed.	1. Proper curing after harvest 2. Store bulbs in dry conditions with good ventilation

DRONES IDENTIFYING PESTS AND DISEASES IN CROPS

- Drones outfitted with multispectral and hyperspectral sensors take images across many wavelengths, extending beyond the visible spectrum. These sensors can identify minor alterations in plant health that are frequently linked to pests or illnesses. The primary instruments in this context comprise:

- **NDVI (Normalized Difference Vegetation Index):** NDVI is utilized to evaluate plant health. Healthy crops have higher reflectance of near-infrared light and lower reflectance of visible light, whereas stressed plants, affected by pests or diseases, demonstrate increased reflectance of visible light and decreased reflectance of near-infrared light. Drones collect this data, facilitating the prompt detection of stress.
- **NDRE (Normalized Difference Red Edge):** This index exhibits heightened sensitivity to variations in plant health within the red-edge spectrum, making it especially adept at identifying early manifestations of disease or nutritional deficits induced by pests.

- Thermal imaging cameras mounted on drones identify temperature fluctuations throughout the area. Pest or disease infestations in plants can induce stress, leading to irregular temperature profiles. For instance:
  - Pests such as aphids or caterpillars can induce stunted growth, leading to temperature discrepancies between stressed and healthy plants.
  - Diseases can induce the plant to deactivate segments of its water transport system, resulting in alterations in leaf temperature, which can be monitored by thermal sensors.
- Drones equipped with high-resolution RGB (red, green, blue) sensors collect photos in the visible light spectrum. These photos can assist in recognizing observable indicators of pests and diseases, including:
  - Leaf darkening resulting from fungal infections or bacterial diseases (e.g., early or late blight).
  - Physical damage caused by insects such as caterpillars or beetles, resulting in discernible perforations in leaves.
  - Webbing produced by pests such as spider mites or caterpillars.
  - Image processing software can examine high-resolution photos to identify anomalies in plant color, shape, and texture.
- LiDAR (Light Detection and Ranging) is not directly employed for pest and disease detection; nonetheless, technology facilitates the generation of intricate 3D maps of crop canopies. LiDAR data can identify areas with unusual growth patterns by evaluating the height, volume, and structure of plants, potentially indicating pest damage or disease prevalence.

COMMON PESTS AND DISEASES IDENTIFIED BY DRONES

Aphids

- **Detection:** Aphid infestations may result in leaf curling and yellowing. Drones outfitted with multispectral or thermal sensors can identify early indicators of these problems by assessing alterations in plant health. The discoloration or drooping linked to aphids can be detected via NDVI and NDRE.

- **Symptoms:** stunted growth, chlorosis of leaves, aphid infestation on the undersides of leaves, and the excretion of sticky honeydew.

Whiteflies

- **Detection:** Drones can recognize whiteflies by locating regions with yellowing or stippling on leaves, indicative of stress resulting from whitefly feeding.
- **Symptoms:** Chlorosis of leaves, black sooty mold, and the presence of whiteflies on the abaxial surfaces of leaves.

Spider Mites

- **Detection:** Spider mites induce a mottled appearance on foliage, identifiable with high-resolution RGB cameras or thermal sensors that emphasize temperature variations.
- **Symptoms:** fine webbing on leaves, yellowing or stippling, and leaf abscission.

Caterpillars

- **Detection:** Drones can identify damage inflicted by caterpillars through observable indicators such as leaf perforations and the presence of frass (caterpillar excrement). Drones can delineate impacted regions utilizing RGB and multispectral sensors.
- **Symptoms:** irregular perforations in leaves, caterpillars consuming crops, and frass present on the ground.

Powdery Mildew

- **Detection:** Drones equipped with multispectral or hyperspectral sensors can identify early indicators of powdery mildew by analyzing alterations in plant reflectance, particularly within the red-edge and near-infrared spectrum.
- **Symptoms:** A white, powdery buildup on leaves and stems, leaf curling, and chlorosis (yellowing) of affected plant tissues.

Late and Early Blight

- **Detection:** Drones outfitted with thermal and multispectral sensors can identify temperature and reflectance variations resulting from late and early blight infections in plants. These illnesses induce abnormal leaf pigmentation and withering, which drones may detect from the aerial perspective.
- Symptoms: Dark, water-soaked sores on foliage, chlorosis, and blackened or decayed fruit.

Rust Diseases

- **Detection:** rust infections, characterized by unique orange, yellow, or red pustules on foliage, can be accomplished utilizing drones equipped with multispectral or high-resolution cameras.
- **Symptoms:** include orange or red pustules on leaves, early leaf abscission, and reduced plant growth.

Fusarium Wilt

- **Detection:** Fusarium wilt impacts the plant's vascular system, resulting in chlorosis and wilting. Drones can recognize variations in plant health via NDVI or hyperspectral sensors, hence detecting stressed vegetation.
- **Symptoms:** Chlorosis of leaves, desiccation, browning of vascular tissue, and plant mortality.

BENEFITS OF DRONE-ASSISTED PEST AND DISEASE DETECTION

- **Early Detection:** Drones facilitate the early detection of pests and illnesses, enabling farmers to mitigate issues before they proliferate and do substantial damage.
- **Precise Mapping:** Drones generate accurate maps of impacted regions, facilitating more focused application of pesticides or fungicides, hence minimizing expenses and environmental repercussions.
- **Real-Time Monitoring:** Drones facilitate continuous surveillance of crop health, enabling prompt interventions for pest or disease incidents.
- **Cost-Effectiveness:** Drones efficiently survey extensive areas, minimizing the time and effort needed for manual scouting. This renders it an economical choice for extensive agricultural operations.
- **Improved Decision-Making:** Data gathered by drones can be amalgamated with other precision agriculture technologies to refine pest management tactics and enhance overall crop vitality.

Challenges in Using Drones for Pest and Disease Detection

- **Data Interpretation:** Analyzing data gathered by drones necessitates specific software and experience. Agriculturalists may require supplementary training to accurately evaluate the results.
- **Weather Conditions:** Drones are susceptible to meteorological factors such as precipitation, wind, and fog, which might affect data accuracy.
- **Initial Investment:** The expense associated with drone technology, sensors, and software may be a challenge for smaller farms; nevertheless, the long-term advantages frequently surpass the early costs.
- **Regulatory Restrictions:** In certain areas, drone operations are forbidden or necessitate special permits, potentially constraining their utilization.

MONITORING STRATEGIES FOR PEST INFESTATIONS AND DISEASE OUTBREAKS

The utilization of agricultural drones for monitoring pest infestations and disease outbreaks is increasingly integral to contemporary precision farming. Drones outfitted with sophisticated sensors and imaging technologies enable farmers to identify pest and disease problems promptly, manage them effectively, and reduce crop damage. This document outlines the monitoring tactics for pest infestations and disease outbreaks with drones.

Pre-Season and Routine Crop Monitoring

- **Objective:** Conduct routine monitoring of crops to evaluate general health and detect potential pest and disease problems prior to their proliferation.
- **Method:** Drones can traverse fields at predetermined intervals to collect data on crop health utilizing multispectral, hyperspectral, and RGB cameras. Pre-season flights may concentrate on evaluating soil health, plant vitality, and identifying initial indications of pests or diseases that are emerging.

NDVI (Normalized Difference Vegetation Index) is utilized to assess overall plant health and identify early indicators of stress, frequently resulting from pests or diseases.

- Thermal sensors to detect regions of stress caused by water or pest damage.
- **Advantages:** Regular monitoring facilitates the early identification of insect activity, disease advancement, or deficiencies that could render crops susceptible to pests and diseases. Early identification of these threats enables farmers to take proactive measures.

Early Detection of Pest and Disease Outbreaks

- **Objective:** Rapidly detect limited pest infestations and disease outbreaks to avert their expansion to broader regions.
- **Method:** Drones equipped with multispectral or hyperspectral imaging sensors are deployed over the field to acquire comprehensive spectral data of the crops. These sensors can detect nuanced alterations in plant health resulting from insect activity or disease prior to the manifestation of visible symptoms.

Tools

- Utilization of NDVI to identify plant stress induced by pests or diseases.
- Red-Edge or NDRE (Normalized Difference Red Edge) is utilized to evaluate plant health with greater accuracy, as diseases such as powdery mildew or blight can modify the reflectance of crops.
- Thermal cameras to detect temperature variations in regions impacted by pests (e.g., aphids or caterpillars) or diseases (e.g., fungal infections).
- **Advantages:** Early diagnosis minimizes the application of pesticides or fungicides, decreases crop loss, and obviates the necessity for indiscriminate treatments. Farmers may implement interventions solely in impacted regions, so reducing expenses and enhancing sustainability.

Targeted Pest and Disease Mapping

- **Objective:** Develop comprehensive maps of pest infestations and disease outbreaks to enable targeted interventions and reduce pesticide application.
- **Method:** Following the identification of possible pest or disease concerns from routine flights, drones can produce high-resolution, georeferenced maps of the impacted regions. These maps identify the locations of concentrated pest populations and the areas experiencing disease proliferation.

Tools:

- High-Resolution RGB Cameras to capture clear images of the infestation or disease symptoms.
- LiDAR for 3D mapping of crop structure, helping to identify areas with irregular growth patterns caused by pests or diseases.
- **Benefits:** These maps help farmers implement precision pest management, directing pesticide or fungicide application to the exact locations that need it, reducing chemical use and environmental impact.

Pest Movement and Spread Monitoring

- **Objective:** Monitor the dissemination and proliferation of pests throughout the field to ascertain the rate of infestation expansion.
- **Method:** Drones may be deployed at regular intervals to monitor the temporal spread of pest populations. Drones can monitor insect movement over various sections of the field by recording photographs at varied seasonal intervals.

Tools:

- Multispectral and RGB Cameras to track visual signs of pest damage.
- Thermal sensors to detect temperature changes in areas with a high concentration of pests, as pest-infested areas tend to have different temperature profiles.
- **Benefits:** By understanding how pests move through fields, farmers can predict which areas are likely to be affected next and take preventive action. This targeted approach helps prevent widespread infestations.

Real-Time Data for Decision Making

- **Objective:** Furnish farmers with prompt feedback to enable timely interventions.
- **Method:**Drones with real-time data transmission capabilities allow farmers to obtain quick updates following each flight. The data can be evaluated locally using onboard software or via cloud-based services.

Tools:

- Cloud-based tools for the integration of drone data with meteorological, pest lifecycle, and agronomic models.
- Farm management software that integrates real-time data from drones with agronomic knowledge to provide actionable insights.
- **Benefits:** Real-time monitoring ensures quick responses to pest infestations and disease outbreaks. Farmers can decide whether to apply pesticides, fungicides, or other treatments, and can adjust their strategies based on current field conditions.

Post-Treatment Monitoring and Evaluation

- **Objective:** Assess the efficacy of pest and disease management strategies post-application.

- **Method:** Following pest control interventions (such as pesticide application or biological management), drones may be employed to surveil the area and evaluate if the issue has been rectified or if additional treatment is necessary.

Tools

- Utilization of multispectral imaging to assess recovery or ongoing harm to crops post-treatment.
- RGB cameras are utilized to detect the physical presence of pests or indications of sickness.
- Thermal sensors to monitor plant temperature recovery after pest or disease management interventions.
- **Benefits:** Post-treatment monitoring guarantees the effective management of the pest or disease problem. It facilitates the refinement of subsequent actions and aids in preventing the excessive application of chemicals.

Pest and Disease Risk Prediction

- **Objective:** Forecast impending pest and disease threats based on prevailing circumstances.
- **Method:** Drones can gather data that, when integrated with meteorological information and agronomic models, can aid in predicting pest and disease threats in the forthcoming weeks. Certain diseases flourish under particular environmental circumstances, such as elevated humidity and precipitation. Drones can monitor agricultural conditions and pinpoint regions with elevated susceptibility to insect infestations or disease outbreaks.

Tools:

- **Weather integration:** Combining drone data with weather forecasts (such as temperature and humidity) allows for predictions on when pests or diseases are most likely to thrive.
- **Field data analytics:** Using past field data and pest lifecycle models to predict when pests or diseases are likely to be present.
- **Benefits:** This predictive approach allows farmers to plan ahead, applying preventative measures or treatments before an outbreak occurs, saving both time and money.

Integration with Other Precision Agriculture Technologies

- **Objective:** Integrate drone-assisted pest and disease surveillance with additional precision agriculture technologies to enhance comprehensive farm management.
- **Method:** Drone data can be amalgamated with additional precision agriculture instruments such as satellite imaging, soil sensors, and variable rate irrigation systems. This facilitates a comprehensive strategy for pest and disease management, guaranteeing that all facets of crop health are considered.

Tools

- Soil moisture sensors to assess situations that may increase crop vulnerability to disease.
- Crop modeling software to simulate and forecast pest or disease dispersion depending on environmental factors.
- **Benefits:** The integration with other technologies enhances the precision of data-driven decisions in the management of pest infestations and disease outbreaks. This method enhances overall agricultural efficiency and sustainability.

Assessment

Assessment involves the analysis of monitoring data (e.g., pest population or damage indicators) to address the inquiry: "Is the pest inflicting damage?"Are the existing preventative or avoidance strategies effective?"Is there a necessity for action?". These inquiries focus on the notion of an action threshold, the juncture at which remedial actions must be implemented to avert pests in a region from inflicting intolerable damage or detriment. The evaluation of insect populations and the formulation of management decisions are fundamental components that distinguish Integrated insect Management (IPM) from alternative pest control systems. The threshold notion is elaborated upon below.

Understanding Thresholds

Prior to implementing any suppression measures, the IPM methodology evaluates if the pest population has surpassed a predetermined action threshold, which signifies the necessity for pest control intervention based on population levels or environmental conditions. Consequently, the identification of a solitary pest or minimal pest populations does not invariably necessitate intervention. It is important to note that thresholds are particular to crops, pests, and meteorological conditions. Certain pests in a field may be below the threshold while others exceed it.

Five considerations are taken into account while establishing action thresholds:

1. **Economic science:** Economic thresholds for numerous pests, predominantly insects, and crops have been meticulously established. A cultivator should refrain from pest management until the population or the resultant damage attains a threshold where the likelihood of economic loss is unequivocal.
2. **Concerns around health and safety:** When the pest poses a health or safety risk, the action threshold is established at a minimal level. If Lyme disease, eastern equine encephalitis, or West Nile virus were endemic in proximity to a playground, and the mosquitoes or ticks that transmit these diseases are present, this would activate the action threshold, irrespective of whether there is one mosquito or one hundred.
3. **Visual considerations:** The degradation of aesthetics prompts aesthetic concerns, such as bird droppings on statues and monuments, or the defoliation of landscape plants and high-value horticultural crops.

4. **Collective sentiment:** Although it may not be immediately apparent, public opinion frequently establishes a notably low action threshold. Regardless of whether the source is societal, cultural, or psychological, certain organisms, such as spiders, are considered pests due to their potential to evoke fear at any degree.
5. **Regulatory obligations:** Specific pests must be managed in accordance with federal, state, or county health standards. Consequently, regulatory standards for action levels are being established. For instance, there is essentially zero tolerance for mice, rats, flies, cockroaches, or any other pests in food establishments. In the event of a declared public health emergency, governmental authorities may impose pest control measures during a rabies or encephalitis outbreak. Building laws and standards may dictate measures to manage termites, rats, or flies on residential or commercial premises as well as public spaces, including parks and schools. In agricultural crops, quarantine zones may exist where pests must be managed to prevent their dissemination to other crops or locations.

Precision agriculture encompasses the utilization of technologies such as remote sensing and geo-referenced crop scouting (to assess crop vigor, quality, and disease), yield monitors (to indicate production variability), and crop and pest modeling. Targeted applications of pest management products are frequently more efficacious than traditional techniques and necessitate reduced pesticide usage. Yield monitors and site-specific crop quality data serve as an assessment of the efficacy of agricultural production inputs and management strategies. A disease predicted to be active by weather monitoring and probable infestation levels can serve as the foundation for variable rate fungicide administration and offer decision support for diverse production strategies, including rootstock selection, fertilization, and irrigation.

- Geographic Information Systems (GIS) are databases that maintain the correlation between acquired data and their geographical locations. A GIS integrates digital mapping, database capabilities, and geographical analysis.
- The Global locating System (GPS) is a United States space-based system that delivers dependable locating, navigation, and timing services globally.

An illustration of a pragmatic application of these devices in Integrated Pest Management (IPM) is as follows.

- The GPS system is utilized to delineate blacklight trap locations in agricultural regions for the surveillance of crop pests. The GPS devices possess an accuracy of within 10 feet.
- The trap counts are gathered and inputted into a spreadsheet/database connected to the GIS program.
- The GIS software is utilized to generate a map of the numbers.

The maps are disseminated to community producers via newsletter, fax, email, or a website.

- Charting the trap counts is advantageous for forecasting the timing and location of pest arrivals. Furthermore, pest maps can be superimposed on meteorological maps to facilitate the prediction of insect migration and the potential for agricultural damage.

- Weather and climate monitoring can anticipate pest infestations or vulnerable life phases. Factors like as humidity and rising degree days can aid in predicting disease outbreaks, such as late blight in potatoes or the crawler stages of scale insects, allowing for the timely application of insecticides to safeguard the plants.
- Weed mapping can be employed to track alterations in weed populations and assess the effectiveness of weed management strategies.
- Spreadsheets can consolidate observations into a single repository. Monitoring data must be documented and encompass details such as date, time, trap placement, pest count, and other relevant environmental factors, in addition to the pest control techniques implemented. Understanding the correlation between locations and data is essential. This process can be conducted with several levels of technological sophistication, ranging from high-tech to low-tech, such as creating a basic map of the farm, urban layout, or other sites with numerical annotations. Data must be examined temporally, spatially, and across other pertinent parameters.

Cameras ought to be utilized often to gather observations in both video and photographic formats. Observations should be timestamped and location-stamped wherever feasible (a smartphone facilitates this process).

Decision Support Tools and Area Wide Pest Monitoring Networks

Although spreadsheets and independent photo collections may suffice for small-scale decision-making, more advantages can be obtained by engagement in regional pest monitoring networks utilizing customized data gathering tools. These systems are often refined for scout-oriented workflows and offer visuals and indicators that distinctly suggest suitable management actions while also providing insights into the experiences of individuals in comparable situations.

Pest and Disease Management

Timely identification and control of pests and diseases are essential to avert substantial crop losses. Drones improve pest and disease management through:

- **Identifying Infestations:** High-resolution images and AI analysis may accurately identify regions impacted by pests, weeds, or diseases, facilitating targeted interventions.
- **Evaluating Treatment Efficacy:** Following treatment applications, drones can assess the success of pest and disease management strategies, facilitating necessary modifications.
- **Reducing Chemical Use:** Drones facilitate the reduction of pesticide and herbicide application by concentrating treatment efforts on specific locations, hence fostering more sustainable agricultural practices.

Figure: 8.1 Identification of Crop Stress

Precise Application of Pest Management

Drones outfitted with spraying mechanisms may accurately administer insecticides and other pest control substances to targeted regions. Drones equipped with GPS and onboard sensors can autonomously traverse fields or designated areas, guaranteeing uniform application of the treatment. This advanced pest control system is engineered to precisely determine the requisite amount of insecticides for infested regions, enhancing safety and economic efficiency.

Furthermore, the speed and efficacy of these unmanned pest management basics are noteworthy. Drones can visit remote or difficult-to-reach locations, such high-rise structures, rooftops, or dense foliage, where pests may exist. They may rapidly traverse extensive regions in a brief period, minimizing the time and labor needed for manual assessments. This efficiency enables pest management professionals to respond promptly to infestations, averting the proliferation of pests and mitigating possible damage.

Oryctes, a robust and precise drone, transforms agricultural spraying for agribusinesses of varying scales. Drones are extensively utilized in precision agriculture management, where predefined trajectories and measurements enhance farm output and management while minimizing manual work and maximizing outcomes. Despite the high costs associated with developing algorithms and correlating ground truth with spectral signatures, the technique becomes user-friendly once foundational investigations are completed.

Aerial Surveillance

Drones outfitted with high-resolution cameras and thermal imaging sensors are capable of performing airborne surveillance across extensive regions. insect management professionals can utilize this technology to detect potential insect infestations, pinpoint breeding sites, and evaluate the severity of the issue. Drones facilitate swift and precise pest identification through the acquisition of real-time photos and movies. Farmers utilize drone monitoring devices to obtain aerial perspectives of the harvest. This outlines information regarding the water system, various soil types, pests, and fungal infestations. Drone-acquired agricultural images include data in both the infrared and visible spectral bands. Various aspects can be extracted from these photographs, offering insights into plant health that are imperceptible to the naked eye. A vital feature of this technology is its ability to consistently check the yield, whether weekly or even hourly. Crop data is often accessible, allowing farmers to implement essential modifications for improved crop management.

Drones can be programmed to routinely survey and monitor designated areas for indications of pests. Through the acquisition of photos and longitudinal data analysis, they can identify preliminary signs of infestations prior to their proliferation. Timely identification facilitates proactive pest management, mitigating substantial harm and diminishing the necessity for rigorous control methods thereafter.

Safety and Environmental Considerations

Drones provide a safer option compared to conventional pest management techniques that may include ascending elevated structures, entering perilous zones, or subjecting persons to toxic substances. Pest management professionals can mitigate dangers to human health and safety through the utilization of drones. This approach diminishes the necessity for physical labor, lessens human exposure to chemicals, and enhances resource utilization by focusing on specific locations instead of administering treatments uniformly throughout an entire field. If one individual can cover 1 hectare per day, a single drone can cover 6 hectares in 10 minutes.

Moreover, drone-assisted targeted pesticide spraying diminishes chemical usage, hence lessening possible harm to the ecosystem and non-target species. Drones can be utilized in nearly all agricultural operations and are regarded as effective instruments for the swift, dependable, and non-invasive identification of field issues.

Unethical pest control practices can lead to disease, pollution, and soil fertility degradation due to the use of various pesticides, insecticides, and fertilizers; however, appropriate pesticide usage can mitigate these effects. Manual spraying by crop scouts across hectares of land will ultimately impact the neurological system and impair the functionality of human body parts that were exposed.

9

Role of Drones in Assessing Soil Moisture Levels and Irrigation Needs

INTRODUCTION OF ASSESSMENT OF SOIL MOISTURE LEVELS IN DRONES

In contemporary agriculture, effective water management is essential for optimizing crop yields while reducing resource consumption. Drones have become a transformative device for evaluating soil moisture levels and identifying irrigation requirements, providing farmers with accurate, real-time data to enhance water utilization. Equipping drones with sophisticated sensors, including thermal, multispectral, and LiDAR, enables them to assess soil conditions over extensive regions, offering insights on moisture fluctuations and pinpointing arid areas in fields. This data-driven methodology not only improves the precision of irrigation schedules but also mitigates water waste, fosters sustainability, and boosts overall crop vitality. Utilizing drones enables farmers to make informed decisions, resulting in more efficient and cost-effective irrigation methods.

Drone soil analysis utilizes drones fitted with specific sensors to gather data on the physical and chemical characteristics of soil. This data may encompass details regarding soil texture, organic matter content, nutrient concentrations, moisture levels, and pH. Various sensors are employed for distinct soil qualities.

Multispectral sensors acquire data across several wavelengths of light, offering insights on plant health and soil properties. Hyperspectral sensors acquire data across numerous tiny spectral bands, yielding enhanced insights into soil composition.

Thermal sensors quantify soil temperature, serving as an indicator of soil moisture content and additional characteristics. The data is subsequently processed and analyzed to generate comprehensive maps of soil variability within a field, offering farmers critical insights to inform their management decisions.

Drone soil analysis utilizes drones fitted with specific sensors to gather data regarding the physical and chemical characteristics of soil. This data may encompass details regarding soil texture, organic matter content, nutrient concentrations, moisture levels, and pH. Various sensors are employed for distinct soil qualities.

- Multispectral sensors acquire data across several wavelengths of light, offering insights on plant health and soil properties. Hyperspectral sensors acquire data across numerous tiny spectral bands, yielding enhanced insights into soil composition.
- Thermal sensors quantify soil temperature, serving as an indicator of soil moisture content and additional characteristics. The data is subsequently processed and analyzed to generate comprehensive maps of soil variability within a field, offering farmers critical insights to inform their management decisions.
- Soil serves as the cornerstone of agricultural productivity, supplying vital nutrients, water, and structural support for plant development. Comprehending soil qualities is essential for maximizing crop yield, guaranteeing effective resource utilization, and reducing environmental impact.
- Soil analysis yields critical insights into the physical and chemical properties of soil, allowing farmers to make informed decisions regarding soil management strategies. This information is crucial for preserving soil health, enhancing crop yields, and guaranteeing the long-term viability of agricultural systems.
- Drones are profoundly influencing the agricultural sector, providing numerous applications beyond soil assessment. These encompass crop surveillance, pest and disease identification, irrigation oversight, and precision application of fertilizers and insecticides.
- Drones offer high-resolution airborne imagery and multispectral data, allowing farmers to gain an extensive perspective of their fields, detect regions of stress or suboptimal performance, and make informed decisions to enhance crop yield.

This data-centric methodology improves efficiency, lowers input expenses, and mitigates environmental effects. Drones serve as an effective instrument for evaluating soil health and variability in soil analysis, facilitating precision management strategies that focus on distinct areas within a field according to their specific soil attributes.

Enhanced Efficiency

Drones facilitate automated data acquisition, encompassing extensive regions rapidly and effectively. This represents a significant benefit for extensive agricultural enterprises where conventional techniques would be too time-consuming.

This expedited data collection enables farmers to promptly address fluctuating soil conditions, like nutrient deficits or moisture stress, and execute appropriate management decisions. This real-time data enables farmers to tackle issues proactively, reducing possible output losses and enhancing resource efficiency. Moreover, the automated aspect of drone data collecting diminishes the necessity for physical work, thereby reallocating human resources to other critical activities.

Cost Reduction

Compared to traditional methods involving manual soil sampling, laboratory analysis, and expert consultation, drone-based soil analysis significantly reduces labor costs and associated expenses. The elimination of manual sampling and the reduced need for extensive laboratory testing contribute to substantial cost savings.

This cost-effectiveness makes precision soil management accessible to a wider range of farmers, including smaller operations that may have previously found traditional methods financially prohibitive. By optimizing input application and reducing waste, drone-based insights further contribute to long-term cost reductions.

Improved Accuracy and Precision

Drones outfitted with sophisticated sensors deliver high-resolution data, detecting minute differences in soil characteristics throughout a field. This degree of specificity facilitates the accurate administration of fertilizers and other inputs, enhancing resource utilization and reducing environmental impact.

Data-Driven Decision Making

The comprehensive data gathered by drones enables farmers to make informed decisions regarding soil management procedures, transitioning from generalist methods to focused strategies customized for individual field circumstances.

Farmers can enhance fertilizer application, irrigation timing, and crop selection by examining soil qualities, moisture levels, and nutrient content, tailored to the distinct characteristics of various sections within their fields. This data-centric methodology results in augmented crop yields, optimized resource utilization, and heightened overall agricultural profitability.

The capacity to display and analyze data via intuitive interfaces enhances farmers' ability to make educated decisions and execute precision management procedures efficiently.

MANUAL SOIL ANALYSIS VS. SOIL ANALYSIS WITH DRONES

Conventional manual soil analysis entails gathering soil samples from multiple sites within a field, dispatching these samples to a laboratory for examination, and awaiting the results. This procedure may be protracted, arduous, and costly.

Conversely, drone analysis provides a significantly swifter and more efficient methodology. Drones can rapidly gather data across extensive areas, offering immediate insights into soil conditions. Manual sampling yields localized data, whereas drone analysis generates a continuous map of soil variability, delivering a more holistic perspective of soil conditions throughout the entire area.

Figure: 9.1 Drone soil Analysis v/s Manual Sail Analysis

DRONE SOIL ANALYSIS

Data Acquisition

Manual soil analysis necessitates the physical collection of samples, which is labor-intensive and time-consuming, particularly for extensive fields. This procedure entails traversing the field and collecting samples at specified intervals or according to visual evaluation.

Drone-based analysis, utilizing remote sensing, swiftly gathers data across extensive regions, markedly diminishing the time and effort needed.

Spatial Coverage

Manual analysis yields data solely from designated sampling locations, potentially failing to reflect the variability throughout the entire area. The restricted sample size may result in interpolation inaccuracies and an inadequate representation of soil conditions. Drone analysis produces continuous maps, offering extensive coverage and detecting nuanced differences in soil parameters throughout the field.

Cost

Manual analysis entails expenses related to labor for sample collection, transportation of samples to a laboratory, and laboratory analysis fees. These expenses can escalate, particularly for extensive fields necessitating many samples.

Drone-based analysis is typically more economical, especially for extensive operations, as it decreases manpower and laboratory costs.

Time to Results

Manual analysis necessitates awaiting laboratory results, which may require days or even weeks, so hindering decision-making. This delay may be crucial, particularly when dealing with time-sensitive matters such as vitamin deficits. Drone analysis delivers near real-time outcomes, enabling farmers to swiftly address fluctuating soil conditions and execute prompt management strategies.

Data Resolution

Manual analysis provides high resolution at specific sampling points, yielding precise information about the soil at those locations. The restricted sample size leads to diminished overall resolution and may fail to represent the complete range of soil variations throughout the field. Drone analysis, through its ongoing mapping capabilities, delivers excellent resolution throughout the entire field, recording nuanced variations in soil parameters.

Data Acquisition

Drones outfitted with selected sensors traverse the field along a predetermined flight trajectory. This flight path is meticulously crafted to provide comprehensive and overlapping coverage of the designated area, reducing data gaps and assuring precise depiction of soil variability.

The drone's altitude and velocity are meticulously regulated to enhance data quality and resolution. Ground control points (GCPs) with established coordinates can enhance the geographic precision of the acquired data.

Data Processing

The unprocessed data gathered by the drone's sensors is subsequently analyzed utilizing sophisticated software. This procedure encompasses multiple stages, such as geometric correction, radiometric calibration, and the elimination of noise and artifacts resulting from air conditions or sensor constraints. The processed data is subsequently transformed into significant information regarding soil qualities, frequently depicted as georeferenced maps or indices that illustrate certain soil features. This processed data serves as the foundation for later analysis and interpretation.

Data Analysis and Interpretation

Specialized software tools evaluate the processed data to provide detailed maps and reports that illustrate soil variations throughout the field.

These visualizations help elucidate regions exhibiting particular nutrient deficits, disparities in soil texture, or variances in moisture content. Agronomists or soil specialists then analyze the data to offer customized recommendations for soil management techniques, like precise fertilizer administration, variable-rate irrigation, or additions to enhance soil health. This analysis links the data to implementable strategies for enhanced agricultural management.

The Benefits of Using Drones for Soil Analysis

The utilization of drones for soil analysis presents multiple benefits compared to conventional techniques, markedly improving the efficiency and accuracy of soil management activities. These advantages enhance crop yields, decrease input expenses, and mitigate environmental effect.

Rapid Data Acquisition

Drones rapidly traverse extensive areas, acquiring data significantly more swiftly than traditional methods. This swift data collection facilitates the quick recognition of soil variability, permitting the immediate execution of management plans. A rapid response time is essential for resolving urgent concerns such as nutritional shortages or moisture stress.

High-Resolution Data

Drones outfitted with sophisticated sensors, such as multispectral or hyperspectral cameras, deliver high-resolution data, detecting minute differences in soil characteristics throughout a field. This comprehensive knowledge facilitates precise management approaches, focusing on specific areas according to their distinct soil characteristics. This degree of specificity facilitates variable-rate application of inputs, enhancing resource use.

Cost-Effectiveness

Drone-assisted soil analysis markedly decreases labor expenses linked to human sampling and laboratory testing. Drones automate data collection, hence reducing the necessity for considerable manual labor. This renders precision soil management more attainable and economical, particularly for bigger agricultural operations.

Improved Accessibility

Drones can readily reach inaccessible locations, including steep inclines, dense foliage, or saturated fields, which may provide challenges or be unfeasible for conventional approaches. This broadens the parameters of soil analysis, facilitating a thorough evaluation of entire areas, irrespective of topographical obstacles.

Reduced Environmental Impact

Drone-based analysis minimizes environmental effect by improving fertilizer and irrigation operations informed by comprehensive soil data. The accurate use of inputs diminishes nutrient runoff into aquatic systems and conserves water, thereby fostering sustainable agriculture. This focused strategy also diminishes the necessity for excessive chemical applications.

The Drawbacks of Using Drones for Soil Analysis

Drone-based soil investigation presents some benefits, although it also encounters specific limits that warrant consideration.

Initial Investment

The procurement of drones, sensors, and related software requires a significant upfront expenditure. The expense of high-quality sensors, especially multispectral or hyperspectral cameras, can be considerable. This may pose a hindrance for smaller farmers or those with constrained financial means. Investigating lease alternatives or service providers may alleviate this initial financial barrier.

Technical Expertise

Operating drones, processing acquired data, and employing analytical software necessitate technical proficiency. Meticulous flight planning, data calibration, and picture processing are needed for precise outcomes. Agriculturalists may require particular expertise or proficient staff to oversee these technological components.

Regulatory Restrictions

Drone activities are governed by regulations that differ based on location and the type of operation. These regulations frequently relate to airspace limitations, license stipulations, and data confidentiality. Agricultural producers must recognize and adhere to all relevant regulations to guarantee lawful and secure operations.

Weather Dependence

Drone operations are vulnerable to meteorological factors, such as precipitation, high winds, or severe temperatures. Inclement weather may restrict flight operations, thereby postponing data acquisition and affecting urgent management decisions. Adaptable flight planning and other data collection methods may be required.

Data Interpretation

Drones yield useful data, although precise interpretation necessitates knowledge in soil science and agronomy. The data acquired by drones must be contextualized and analyzed in conjunction with elements such as crop kind, climate, and previous soil data. Misinterpretation may result in unproductive or detrimental management decisions. Collaboration with agronomists or soil specialists can guarantee accurate data analysis and interpretation.

OPTIMIZING IRRIGATION SCHEDULING WITH DRONE DATA

Drones outfitted with remote sensing technologies can deliver high-resolution spatial and temporal data regarding soil moisture variability throughout a farm. This data can optimize irrigation scheduling, resulting in water conservation, enhanced crop yields, and diminished environmental impact.

Water is a limited resource, and its optimal utilization is essential for sustainable agriculture. Deep learning models use thermal and multi-spectral drone data to evaluate soil moisture levels and detect water-stressed regions. Principal advantages encompass:

- **Enhanced Irrigation:** Supplying data to modify irrigation schedules according to current moisture levels.
- **Leak Detection:** Identifying regions of excessive irrigation or pipeline leaks.
- **Water Conservation:** Directing water only to regions requiring it, hence minimizing waste.

These apps enhance crop health and promote sustainable agricultural practices through resource conservation.

Deep learning and drone imaging facilitate the evaluation of plant health and the optimization of irrigation, providing farmers with practical information that enhance production, efficiency, and sustainability in contemporary agriculture.

Figure: 9.2 Use of Drone in Modern Agricultural for soil Analysis

Effective irrigation and water management are essential for sustainable agriculture, and drones have surfaced as potent instruments in this field. Drones are transforming agricultural practices by collecting real-time data and delivering accurate aerial insights, enabling farmers to monitor and optimize water usage in their fields.

Mapping and Monitoring Soil Moisture Levels

Drones outfitted with multispectral or thermal imaging cameras can capture precise data regarding the soil's moisture content. By aerial surveying of fields, they can identify variations in soil moisture across extensive regions that may be challenging or labor-intensive to evaluate from the ground. These high-resolution photos empower farmers to:

- Determine regions that are excessively irrigated or insufficiently irrigated.
- Modify irrigation systems to allocate water with more precision.
- Mitigate excessive water usage and reduce the risk of waterlogging or crop stress due to dehydration.

Drones provide the creation of precise irrigation maps customized to the specific requirements of different sections of the field.

Early Detection of Irrigation System Failures

Traditional irrigation systems sometimes suffer from inefficiencies caused by leaks, blockages, or mechanical failures. Drones can traverse irrigation lines, identifying problems such as uneven water distribution, obstructed sprinklers, or pipe leakage. Infrared and thermal sensors identify temperature anomalies indicative of areas where water distribution is inadequate. Prompt identification mitigates water wastage and ensures timely repairs, thereby improving irrigation efficiency.

Precision in Variable Rate Irrigation (VRI)

Drones enable the execution of Variable Rate Irrigation (VRI), a contemporary irrigation method that modulates water application rates according to the specific requirements of various sections of a field. Utilizing drone-generated data, farmers can:

- Segment farmland into zones necessitating differentiated irrigation fees..
- Calibrate sprinklers or drip systems to deliver the appropriate volume of water in each sector.
- This precise method reduces water wastage and guarantees optimal hydration for flowers, resulting in healthier plants and increased yields.

Water Stress Detection in Plants

Drones equipped with advanced imaging technologies, such as Near Infrared (NIR) or hyperspectral cameras, can detect water stress in plants. These cameras capture data beyond the visible spectrum, exposing early indicators of plant stress caused by inadequate or excessive irrigation. By analyzing these facts, farmers can modify irrigation schedules prior to water scarcity impacting crop yield. Proactive water management utilizing drones enhances plant vitality and reduces losses associated with water-related issues.

Drones to Assess Soil Moisture and Irrigation Needs

Drones outfitted with sophisticated sensors and imaging technology can evaluate soil moisture levels, an essential factor for efficient irrigation management. Drones can employ sensors like as thermal and multispectral cameras to collect data on plant stress, evapotranspiration rates, and soil moisture levels. This information allows farmers to fully comprehend the water needs of their crops and make informed decisions about irrigation schedule.

Enhancing irrigation scheduling utilizing drone data is an effective strategy in contemporary agriculture that guarantees crops have the appropriate quantity of water at the optimal moment. Farmers can utilize drones outfitted with sophisticated sensors and imaging technology to monitor field conditions in real-time, evaluate soil moisture levels, and make informed decisions to enhance irrigation efficiency. This results in water conservation, financial savings, and enhanced agricultural output. This document provides a comprehensive examination of the utilization of drone data for the optimization of irrigation scheduling.

Assessing Soil Moisture Levels

- **Objective:** Assess the present soil moisture levels in various sections of the field to pinpoint areas requiring irrigation.
- **Method:** Drones outfitted with multispectral and thermal sensors acquire real-time data regarding crop health and soil moisture.

- **Thermal Imaging:** Drones utilize thermal sensors to identify temperature fluctuations in soil and vegetation. Arid soil generally exhibits a greater surface temperature compared to hydrated soil, enabling the drone to identify regions requiring additional irrigation.
- **Multispectral Imaging:** Multispectral Imaging: By acquiring data across several wavelengths, including near-infrared, multispectral sensors offer insights into plant health and stress levels. Plants experiencing water stress display distinct reflectance patterns that can be discovered early with the use of drones.
- **Benefits:** This data assists farmers in pinpointing regions of the field that are either under-irrigated or over-irrigated, facilitating more accurate water management.

Identifying Crop Water Stress

- **Objective:** Identify preliminary indicators of agricultural water stress prior to their manifestation to the unaided eye.
- **Method:** Drones equipped with NDVI (Normalized Difference Vegetation Index) and NDRE (Normalized Difference Red Edge) sensors are particularly effective in detecting plant stress associated with water scarcity.
- **NDVI:** A high NDVI number signifies robust plants, whereas a low NDVI value indicates stress, potentially resulting from insufficient water. Drones utilizing this indicator can identify stressed regions of the land that may require additional watering.
- **NDRE:** This indicator is particularly responsive to plant vitality, encompassing water stress, and is optimal for identifying initial indications of wilting or desiccation.
- **Benefits:** Early identification of water stress allows farmers to take action before to substantial crop damage, hence averting output loss.

Creating High-Resolution Field Maps for Irrigation

- **Objective:** Develop accurate irrigation maps that provide the specific locations and timings for irrigation requirements.
- **Method:** Drones can be utilized to survey agricultural areas, generating high-resolution maps that depict soil moisture levels and crop vitality. These maps help facilitate irrigation scheduling and determine which locations necessitate increased or decreased water supply.
- **Georeferenced Imagery:** Drones collect georeferenced data, offering accurate location-specific insights regarding soil moisture and crop health, which can be integrated into irrigation management systems.
- **3D Mapping:** Drones can generate 3D topographical maps of the terrain utilizing LiDAR sensors, facilitating the comprehension of elevation variations that may influence water distribution.
- **Benefits:** These comprehensive maps provide variable rate irrigation (VRI), enabling distinct portions of the field to receive tailored water allocations according to their specific requirements.

Real-Time Monitoring and Adaptive Irrigation

- **Objective:** Persistently assess field conditions and modify irrigation schedules in real-time.
- **Method:** Drones may be employed to systematically observe the field at regular intervals to monitor variations in soil moisture, plant vitality, and environmental parameters (including temperature and humidity). This real-time data can be incorporated into irrigation control systems to automatically modify irrigation schedules according to the latest field conditions.
- **Integration with Irrigation Systems:** The incorporation of drone data into automated irrigation systems or intelligent irrigation controllers enables the adjustment of watering schedules according to real-time field data. For instance, if the drone detects an area of the field with insufficient moisture, the irrigation system can be activated to supply water to that particular region.
- **Benefits:** The continuous feedback loop guarantees that crops obtain the ideal quantity of water during the growth season, enhancing water use efficiency and minimizing waste.

Identifying Irrigation Inefficiencies

- **Objective:** Identify regions within the field where irrigation systems exhibit suboptimal performance, leading to irregular water distribution.
- **Method:** Drones outfitted with thermal sensors can identify temperature variations resulting from irregular watering. Regions with inadequate irrigation will see elevated temperatures, whilst excessively irrigated areas will exhibit lower temperatures.
- **Water Distribution Mapping:** Drones can identify overwatered areas in the field by detecting runoff or pooling water through thermal or high-resolution imagery.
- **Benefits:** Recognizing ineffective irrigation systems enables farmers to undertake essential repairs, modifications, or replacements to guarantee uniform and efficient water distribution.

Irrigation Scheduling Based on Weather Forecasts and Crop Stage

- **Objective:** Modify irrigation schedules in accordance with meteorological predictions, crop developmental phases, and soil moisture content.
- **Method:** Drones can deliver real-time information on soil moisture and crop health, which, when integrated with meteorological data (such as impending precipitation or temperature fluctuations), facilitates more precise irrigation scheduling.
- **Weather Integration:** Drone data can be utilized in conjunction with local weather forecasts to optimize watering practices. For instance, if precipitation is forecasted in the coming days, irrigation may be postponed to prevent excessive watering.
- **Crop Growth Stages:** Drones can evaluate the growth stage of crops, which is essential for estimating irrigation requirements. Plants at various developmental stages necessitate varying quantities of water.

- **Benefits:** The advantages of this dynamic scheduling include a reduction in water use by considering both existing field conditions and forecasted weather patterns, resulting in substantial water conservation and financial savings.

Enhancing Water Use Efficiency and Sustainability

- **Objective:** Advocate for sustainable water utilization by enhancing irrigation timetables.
- **Method:** Employing drones for the constant surveillance of soil moisture, crop vitality, and environmental parameters enables farmers to prevent both over-irrigation and under-irrigation, resulting in enhanced water efficiency. Drones can assist in pinpointing spots where irrigation is superfluous, such areas that are inherently more fruitful or possess greater moisture retention.
- **Precision Irrigation:** Drones enable farmers to execute precision irrigation, delivering the appropriate quantity of water to designated regions according to real-time data. This minimizes water waste, guarantees proper crop development, and enhances long-term sustainability.
- **Benefits:** Enhancing irrigation schedules utilizing drone data conserves water and mitigates the environmental impact of agricultural activities. It promotes sustainable agriculture by mitigating water runoff, soil erosion, and waterlogging.

Post-Irrigation Assessment

- **Objective:** Assess the efficacy of irrigation and guarantee consistent water distribution.
- **Method:** Following irrigation, drones may be utilized to evaluate the distribution and efficacy of the water application. Thermal sensors can pinpoint regions that remain dry or excessively wet, but multispectral sensors can identify places where vegetation has recuperated from hydric stress.
- **Follow-Up Surveys:** Drones can be utilized for follow-up surveys to evaluate the long-term effects of irrigation by monitoring alterations in plant health, soil moisture, and crop development.
- **Benefits:** The advantages of post-irrigation assessment enable farmers to modify subsequent irrigation cycles, hence enhancing water use efficiency and crop vitality.

WATER RESOURCE MANAGEMENT AND CONSERVATION USING DRONE TECHNOLOGY

Drones are essential for mapping and monitoring water resources in agricultural environments. They can obtain high-resolution aerial imagery and generate intricate maps of reservoirs, ponds, and other aquatic features on the farm. These maps offer data on water levels, surface area, and possible sources of water depletion, assisting farmers in the efficient management of their water resources. By monitoring water sources, farmers can detect concerns such as leaks or excessive evaporation and implement corrective actions swiftly.

Effective water management is vital for optimizing crop productivity. Drones significantly contribute to these domains by:

- **Surveillance of Irrigation Systems:** Drones equipped with cameras may identify arid areas resulting from water stress in crops and detect leaks or obstructions in irrigation systems, facilitating prompt repairs and modifications.
- **Planning Irrigation Schedules:** Drones facilitate the optimization of irrigation schedules by delivering real-time data on soil moisture and crop water requirements, so assuring effective water utilization.

Drone for Water Conservation and Improved Crop Health

Drone-assisted irrigation provides numerous advantages, such as water conservation and enhanced crop vitality. By accurately detecting regions with differing moisture levels using drone data analysis, farmers may execute targeted irrigation techniques. This focused strategy diminishes water waste, lowers energy usage, and guarantees that crops obtain the appropriate quantity of water at the optimal moment. Consequently, water resources are preserved, and the overall sustainability of agricultural activities is improved. Furthermore, drone-assisted irrigation enhances crop health. Drones can identify early indicators of water deficit or surplus by monitoring crop stress levels and soil moisture content. This information allows farmers to rapidly modify irrigation operations, averting both under-irrigation and over-irrigation, which can result in diminished crop yields or disease proliferation. As a result, crop vitality is enhanced, resulting in increased yields and superior-quality produce.

Data-Driven Water Conservation

Efficient water utilization is crucial in regions prone to drought. Drone analytics assist farmers in making data-driven decisions to conserve water while maintaining crop health. Drones aid farmers in assessing the requisite water quantity and optimizing future irrigation techniques by analyzing ancient statistics with contemporary field conditions. Drones significantly contribute to reducing water consumption by accurately mapping water requirements, aiding farms in adhering to water regulations, and preserving natural resources.

Assessing Field Topography for Optimal Water Flow

Irrigation involves not only supplying water to crops but also ensuring proper flow and drainage. Drones equipped with LiDAR or photogrammetry technology can generate unique topographical maps of landscapes. These maps assist farmers in understanding water flow patterns and drainage systems. Through the examination of slope, elevation, and terrain characteristics, farmers can design irrigation systems that mitigate erosion, runoff, and water accumulation in low-lying areas.

The Advantages of Drone-Assisted Irrigation

Efficiency and Cost Savings

Drones enhance irrigation efficiency by minimizing labor efforts and facilitating accurate water distribution. This results in significant cost reductions in water consumption, electricity, and labor. Drones are automating several aspects of water management, enabling farmers to monitor their resources across diverse areas of agricultural productivity.

Sustainability

In light of escalating environmental concerns, drone-assisted irrigation facilitates sustainable agriculture. Drones aid farms in minimizing water waste and enhancing resource management, so promoting sustainability, lowering environmental impact, and increasing yields.

Real-Time Monitoring and Scalability

The capacity to monitor vegetation and water structures in real-time provides farmers with a strategic advantage, enabling immediate responses to emerging challenges. Furthermore, the drone era may rapidly expand to encompass extensive areas, rendering it ideal for large-scale agricultural operations and places with limited water resources.

Challenges and Considerations

Although drones offer significant advantages in enhancing irrigation, certain challenges must be acknowledged:

- **Initial Expenses:** The investment in drone technology and education can be exorbitant, particularly for small-scale farms. Nonetheless, enduring financial savings from water efficiency and agricultural productivity might counterbalance those expenses.
- **Data Interpretation:** Drone recordings must be effectively processed and analyzed to yield utility. Numerous farmers may also depend on software platforms or third-party service providers to analyze the data.
- **Regulatory Concerns:** In many regions, stringent regulations around drone usage may further constrain their capabilities, particularly for extended flights over agricultural landscapes.

The Future of Drone Irrigation

As the drone era progresses, we should expect increasingly sophisticated instruments for irrigation and water control. Artificial intelligence and knowledge acquisition will enhance drones' capacity for autonomous decision-making, anticipate water requirements, and improve irrigation in real-time. Integration with other smart farming equipment, such as IoT sensors and automated irrigation systems, will yield entirely self-sustaining water management systems.

Drones are facilitating a more sustainable and efficient future in agriculture by reducing water waste and improving irrigation procedures. In a context where every drop is crucial, drone technology is rapidly becoming indispensable for farmers aiming to optimize water utilization, conserve resources, and enhance crop yields.

Drones are transforming agricultural water management. Drones facilitate educated, effective, and sustainable water management decisions for farmers by offering real-time data on soil moisture, identifying irrigation problems, and supporting precision irrigation techniques such as Variable Rate Irrigation (VRI). With the increasing global population and the escalating unpredictability of climate, the implementation of drone technology in irrigation is essential for maintaining food security and promoting sustainable agricultural practices.

10

Routine Maintenance Procedures for Agricultural Drones

INTRODUCTION TO ROUTINE MAINTENANCE PROCEDURES FOR AGRICULTURAL DRONES

Regular maintenance of agricultural drones is crucial for their best performance and durability, as these sophisticated devices are integral to contemporary agriculture. Agricultural drones endure severe climatic conditions, such as dust, wetness, and fluctuating temperatures, which might impact their performance. Routine maintenance activities, including cleaning, inspecting, calibrating sensors, and assessing component wear, mitigate expensive downtime and guarantee that drones consistently provide accurate data for applications such as crop monitoring, irrigation management, and pest control.

By following a regular maintenance schedule, farmers may optimize the dependability and effectiveness of their drone technology, so improving production and reducing the likelihood of malfunctions during essential activities.

UAV systems (drones) necessitate routine maintenance inspections to verify optimal functionality. They are susceptible to deterioration from prolonged usage, particularly when utilized within a bigger fleet.

This checklist may be utilized as frequently as necessary for routine maintenance inspections; however, it is more efficacious to conduct maintenance on a per-flight basis rather than at fixed intervals. Scheduling will more precisely indicate the likelihood of the drone requiring maintenance.

You likely already possess established procedures for determining when and how to conduct maintenance. Verify all items on this list to confirm their operational status and facilitate any required repairs or replacements prior to the next flight.

Figure: 10.1 Remove mud and Grime

Clean Chassis of Mud And Dirt

Your drone will progressively gather layers of detritus and substances as a result of flight, transportation, and storage stresses. The entire device must be cleaned regularly. Most filth and accumulation can be removed with an air duster, a microfiber cloth, and a small amount of water. More comprehensive cleaning techniques may be required for stubborn stains; nevertheless, exercise caution while handling sensitive electronic components.

Inspect Chassis for Cracks

Examine the chassis meticulously for any damage. Consistently do these inspections and ensure all components are maintained as required. Neglected, even the minutest fissure can lead to significant problems. Thoroughly inspect each component during routine maintenance and document any damage seen.

Check for Loose Screws

Ensure that all components are securely affixed. Certain components will likely require removal and replacement during maintenance or repairs. Examine the fastenings and bolts in this context, including those in older components. Refrain from excessive tightness. Joint damage may occur due to excessive strain imposed on them as a consequence of this. The joints must be tightened with moderate force to ensure they are snug and secure.

Check Propellers for Damage

The propellers are among the more sensitive components of your system. They must undergo meticulous inspection for fractures to verify they are secure.

Irrespective of the seamlessness of your prior flight, a meticulous inspection remains important. The last thing you desire is for your equipment to malfunction mid-flight owing to a minor crack you overlooked during routine maintenance. Propeller damage may be unnoticed until it reaches a critical state. Substitute defective components and persist in rigorously testing the motors.

Check Propellers are Free-spinning

Ensure the battery pack is entirely disconnected prior to conducting this check. Examine for any supplementary impediments that may hinder the complete range of motion by sequentially turning each propeller. The propeller should ideally rotate seamlessly and without obstruction. Should the propellers encounter difficulties in rotating freely, consult your unit's handbook for guidance on removing the propellers and meticulously cleaning each component of the motor-propeller module.

Check Motors for Debris and Obstructions

Debris, such as grit, mud, or biological matter, may impede the motor chamber by being lodged in the apertures surrounding the top of the armature. Ensure that the area between the propeller and engine is free of impediments, and clean the casing. It is advisable to detach the propellers to obtain direct access to the motor compartment and meticulously eliminate all extraneous things. To eliminate obstinate debris, it may be need to disassemble the drone further; in this instance, see your model's instruction manual for the proper disassembly and reassembly procedures.

Check the State of the Wiring and Solder Joints

Given its susceptibility to damage, you should prioritize inspecting any exposed wiring in your device. Ensure that all components are secure and inspect for visibly worn or frayed areas. It is necessary to examine the internal wiring by accessing the chassis. Prior to soldering any loose connections, refer to the unit's instructions. Improperly soldering a wire or neglecting essential measures while handling internal components may result in the destruction or impairment of the circuitry.

Check Unit Camera is Clean

Clean the camera lens and remove any dust or debris from the body and frame. This component must be meticulously cleaned during routine inspections, as bug residue and other environmental conditions can lead to considerable dirt accumulation.

Check the Landing Gear Condition

Ensure that all rubber shock absorbers are intact and that none of the unit's legs or feet exhibit bending or cracking. The drone was unable to land safely without this module. All damaged or missing components must be procured from the manufacturer and reinstalled.

Inspect Antennae

The antennae are responsible for maintaining wireless communication between the mobile drone unit and the ground control station. A catastrophic loss of control may occur due to impaired connectivity from damaged or incorrectly installed antennae. Ensure that the screws securing each antenna to the device are tightened.

DIAGNOSING AND TROUBLESHOOTING COMMON ISSUES (MOTOR FAILURE, GPS SIGNAL LOSS, SENSOR CALIBRATION, ETC.)

Diagnosing and resolving prevalent issues with agricultural drones is essential for guaranteeing optimal performance and reducing downtime. Issues such as motor malfunction, GPS signal degradation, and sensor calibration inaccuracies might impede operations and compromise data collecting precision. This is a handbook to assist in troubleshooting some prevalent issues:

Motor Failure

Symptoms

- The drone exhibits erratic behavior or struggles to achieve lift-off.
- Anomalous auditory or vibrational phenomena emanating from the drone.
- The drone exhibits difficulty in sustaining stable flying or demonstrates unpredictable behavior.

Possible Causes

- Motor deterioration or impairment.
- Loose or compromised motor connections.
- Obstruction of the motor or propeller by debris.
- Overheating or electrical failures in the motor.

Troubleshooting Steps

- **Examine for physical damage:** Assess each motor for observable indications of deterioration, fractures, or excessive heat exposure. Examine for indications of physical impairment to the motor or propeller.
- **Clean the motors:** Eliminate any dirt, debris, or botanical matter that may have gathered around the motors and propellers. Utilize pressurized air to purge the vicinity.
- **Inspect motor connections:** Verify that all cables and connections to the motors are secure. Frayed or compromised wires may result in motor malfunction.
- **Evaluate each motor separately:** If the problem continues, assess each motor independently to detect any defective units.
- **Substitute defective motors:** If a motor is identified as malfunctioning, replace it with an appropriate component.

GPS Signal Loss

Symptoms

- The drone exhibits erratic behavior or struggles to maintain a stable position.
- The drone experiences a loss of positional accuracy on the map or encounters challenges in adhering to waypoints.
- Warnings such as "No GPS" displayed on the controller.

Possible Causes

- Weak or obscured GPS signal.
- Interference from adjacent structures (e.g., tall edifices, vegetation).
- Defective GPS module or antenna.
- Improper software configurations.

Troubleshooting Steps

- **Assess the Drone's Surroundings:** Confirm that the drone is operating in an unobstructed region with limited interference from substantial structures such as buildings or trees, which may obstruct or diminish the GPS signal.
- **Reboot the Drone:** Deactivate the drone and power it on again to reset the GPS system.
- **Examine the GPS Module:** Assess for any loose or compromised cabling linking the GPS module to the drone. Should the GPS antenna or module sustain damage, replacement may be necessary.
- **Upgrade Firmware:** Verify that the drone's firmware is current. GPS-related malfunctions may occasionally be rectified by software updates.
- **Conduct a GPS Signal Assessment:** In certain instances, drones permit users to evaluate the intensity of GPS signals. Assess the signal intensity on the controller and juxtapose it with the requisite values for stable flying.

Sensor Calibration Issues

Symptoms

- Erroneous or inconsistent data readings from the sensors (e.g., crop vitality, soil moisture levels).
- The drone transmits erroneous or absent data during operations.
- The drone's navigation system malfunctions (e.g., deviating or failing to adhere to designated flight paths accurately).

Possible Causes

- Sensors are inadequately calibrated or have deviated from their ideal configurations.
- Environmental variables (e.g., temperature, humidity, or interference) influencing sensor efficacy.
- Firmware or software malfunctions impacting sensor readings.

Troubleshooting Steps

- **Recalibrate sensors:** Adhere to the manufacturer's guidelines to recalibrate the drone's sensors, encompassing the camera, multispectral sensors, and thermal sensors. Calibration often entails operating the drone within a regulated environment or employing a calibration instrument to verify the accurate alignment of the sensors.
- **Examine sensor connections:** Assess the wiring and connectors associated with the sensors. Loose or compromised connectors may result in data complications.
- **Remove contaminants from the sensors:** Accumulations of dirt, dust, or moisture on the sensors, particularly the camera or lenses, may disrupt readings. Utilize a gentle microfiber cloth to cleanse the sensor lenses and verify they are devoid of any obstructions.
- **Verify software updates:** Occasionally, sensor calibration problems are associated with obsolete firmware or software. Verify that the drone's software is current and that all requisite calibration algorithms are installed.
- **Environmental factors:** Verify that the drone operates under conditions favorable for accurate sensor readings. Severe temperatures, humidity, or environmental factors may influence the precision of specific sensors.

GPS Drift and Altitude Instability

Symptoms

- The drone deviates from its designated flying trajectory, even as it is meant to remain stationary.
- The drone experiences difficulty in sustaining a consistent altitude, resulting in erratic ascents and descents.

Possible Causes

- GPS disruption or inadequate satellite connectivity.
- Improper calibration of the Inertial Measurement Unit (IMU).
- Problems with the compass or gyroscope.

Troubleshooting Steps

- **Assess GPS interference:** Operate the drone in an unobstructed location, apart from substantial metallic edifices, vegetation, or electrical apparatus that could disrupt the GPS signal.
- **Recalibrate the IMU:** The IMU is essential for stabilizing the drone's orientation. A calibration problem may result in drift or erratic flight behavior. Consult the drone's instructions to recalibrate the Inertial Measurement Unit (IMU).
- **Compass Calibration:** Verify the accurate calibration of the compass. A compass inaccuracy may result in the drone exhibiting drift or improper altitude

regulation. Execute a compass calibration according to the manufacturer's specifications.

- **Examine for Magnetic Interference:** Verify the absence of surrounding magnetic sources (e.g., power lines or metallic structures) that could disrupt the compass.

Battery-Related Issues

Symptoms

- Brief flight durations or unforeseen battery exhaustion.
- The drone either fails to launch or ceases operation during flight.
- The battery is bloated, compromised, or fails to charge.

Possible Causes

- Battery degradation or impairment.
- Improper billing protocols.
- Loose or compromised battery terminals.

Troubleshooting Steps

- **Examine the Battery:** Look for indications of swelling, leaking, or physical impairment to the battery. Should any of these indicators be evident, the battery must be changed without delay.
- **Inspect Battery Connectors:** Verify that the battery is correctly positioned and that the connectors are clean and firmly attached.
- **Proper Charging Procedures:** Adhere to the manufacturer's guidelines for battery charging. Utilize the appropriate charger and refrain from overcharging or undercharging the battery.
- **Examine Using an Alternative Battery:** Should the issue continue, utilize a different battery to ascertain whether the problem pertains to the battery or the drone itself.

Communication Loss Between Drone and Controller

Symptoms

- Loss of control or delayed reaction to stimuli.
- The drone exhibits unresponsiveness when in flight.
- "Signal Lost" or "Communication Error" notification on the controller.

Possible Causes

- Weak or disrupted signal between the drone and the controller.
- Interference from additional wireless devices.
- Defective communication apparatus in the drone or controller.

Troubleshooting Steps

- **Verify Signal Range:** Confirm that you are inside the drone's signal range. Exceeding the operational range of the controller may result in a loss of communication.
- **Frequency Modification:** If interference from proximate devices (e.g., Wi-Fi, other drones) is anticipated, alter the communication frequency on the controller.
- **Reboot Both Devices:** Power cycle the drone and the controller to restore the connection.
- **Examine Antennae:** Assess the drone and controller antennas for any damage. Malfunctioning or misaligned antennas may result in communication problems.

Software Glitches or Errors

Symptoms

- The drone exhibits erratic behavior or does not adhere to flight plans.
- Deficiencies or malfunctions in the flight application or flight control software.
- The drone exhibits unresponsiveness to commands or experiences delayed processing times.

Possible Causes

- Obsolete software or firmware.
- Software defects or malfunctions.
- Discrepancies with external applications or tools.

Troubleshooting Steps

- **Upgrade Software:** Confirm that the drone's firmware and the controller's software are current. Updates may rectify errors and enhance stability.
- **Reboot Devices:** Reboot the drone, controller, and mobile device operating the flight control software to eliminate transient software anomalies.
- **Restore the Drone to Factory Settings:** Should software issues continue, restoring the drone to factory settings and reinstalling the software may rectify the problem.
- **Verify Application Compatibility:** If utilizing third-party software, confirm its compatibility with your drone's firmware and hardware.

Software/Firmware Update

Updating the firmware on your drone ensures that all components are up to date and functioning properly. Regularly updating the system mitigates the risk of security vulnerabilities and promotes optimal functionality. The firmware can be effortlessly updated by connecting the drone to a computer using a USB or comparable cable and manually installing the device-specific software. Firmware can also be updated wirelessly via an application or the control station interface. Utilize the most appropriate method to upgrade the system firmware prior to proceeding to the next procedure. Always refer to the documentation specific to your item to understand the procedure for updating the firmware. Various approaches will be employed based on the model and manufacturer.

Custom software is vital for the efficient functioning of your control station. This can be modified in a manner akin to the drone unit, and your control station will generally possess integrated mechanisms. Control programs for mobile devices, such as smartphones or tablets, will update automatically unless directed differently; nonetheless, it is advisable to verify this.

Employ the most effective method to upgrade the software of your control station, if required. Comparable updating techniques will be included into bigger units, such as portable laptop computers operating proprietary software. Upon completion of all updates, you will be fully prepared. The device must complete several straightforward testing protocols prior to its forthcoming flight.

BATTERY MANAGEMENT AND CARE

The docking station for each battery pack is often overlooked. This component is often the primary source of battery-related issues and typically remains unmaintained for extended periods. Ensure that the voltage shown by the dock is compatible with that of the battery pack and conduct a visual inspection.

Subsequently, do a thorough examination of each battery pack. Bulges or abnormalities signify leakage, necessitating the rapid replacement of the damaged packs. Regular maintenance must be conducted as necessary to sustain and extend the longevity of battery packs. Inspecting and substituting active packs and spares as required.

- Examine drone batteries for any signs of damage.
- Examine controller batteries for any damage.
- Examine supplementary component batteries for damage.
- Substitute defective batteries (if required)

Prior to completing this task, verify that each pack retains a minimum of 75% of its charge following inspection and, if required, replacement. Batteries must not be overcharged or permitted to discharge entirely. The charge should optimally range from 30% to 90%.

How to Maintain the Agriculture Drone Battery Routine

- Check the battery regularly. Handles, wires, plugs, and check the damage to the appearance. Deformation, corrosion, discoloration, ragged surface, and whether the plug is too loose between the plug and the aircraft.
- After each operation, you must wipe the appearance of the battery and plug with a dry cloth to ensure that no pesticide residue to avoid battery corrosion.

The battery temperature is high after the flight and needs to be charged once the battery temperature drops below 40°C (the recommended temperature range for battery charging is 5°C to 40°C).

- Charging the battery after its exterior temperature has cooled can considerably improve the battery's service life.
- Summer temperature is relatively high, it is preferable not to expose the battery to the sun.
- In winter, after the battery is depleted, the battery should be utilized useful insulation measures (such as the use of insulation boxes) to ensure that the battery temperature is maintained at 5 °C or above, and the battery life is considerably shortened in low temperature environment.

Figure: 10.2 A Drone Utilizing Battery

Case of Emergency

When the battery is on fire during charging, first of all, you should cut off the power of the equipment, use asbestos gloves or clips to remove the burned battery, put it on the ground or in a bucket of fire sand, cover the burned lithium battery with an asbestos blanket, and put fire sand over the asbestos blanket to block the air. If you need to scrap the depleted battery, please thoroughly soak the farm drone battery in salt water for more than 72 hours to ensure that it is entirely discharged, and then dry it before doing the scrap procedure. Do not use dry powder extinguishers to extinguish open fires because for solid metal chemical fires, a substantial volume of dry powder is necessary to cover the fire, which causes corrosive pollution to the equipment. The use of carbon dioxide to extinguish fire will not contaminate the environment and will not corrode the equipment, but it can only achieve the effect of rapidly reducing the flames. It needs to be used combined with sandstone and asbestos blankets. Blocking and suffocation is the best technique to cope with the combustion of the drone batteries.

Agricultural drones are costly. Upon discovering a fire, it is imperative to promptly notify expert assistance to mitigate property loss and personal injury. This is an introduction to the daily maintenance and emergency disposal techniques for agricultural drone batteries. Acquiring expertise in the daily maintenance protocols and emergency disposal techniques for agricultural drone batteries is an essential responsibility for every agricultural drone operator. It can significantly enhance the cycle life of the agricultural drone's battery, hence optimizing the drone's performance.

11

Safety Protocols for Drone Operations in Agricultural Settings

INTRODUCTION TO SAFETY PROTOCOLS FOR DRONE OPERATIONS

Drones in agriculture have transformed how farmers oversee and cultivate crops, enhance efficiency, and optimize resources. Nonetheless, like any sophisticated technology, the utilization of drones in agricultural contexts necessitates compliance with stringent safety measures to safeguard the operator, the surrounding environment, and the equipment. The size, power, and complexity of agricultural drones entail inherent operational hazards, including possible accidents, collisions, and interference with other equipment or humans. Implementing safety measures not only prevents accidents but also guarantees the accuracy and reliability of acquired data, hence optimizing the advantages of drone technology in agriculture. By complying with safety regulations, such as conducting thorough pre-flight inspections, establishing secure flight areas, and implementing emergency response protocols, farmers can mitigate risks while maximizing the benefits of drones in agricultural management.

Implementing safety guidelines for drone operations in agricultural environments is crucial to safeguard both operators and the ecosystem while optimizing technological efficacy. Drones serve as potent instruments for activities such crop monitoring, spraying, and soil health evaluation; nevertheless, inadequate management or noncompliance with safety protocols may result in accidents, equipment damage, and data integrity issues. The following are essential safety protocols to adhere to when operating drones in agricultural environments:

Pre-Flight Safety Checks

Equipment Inspection

- **Drone:** Inspect for any apparent damage, fractures, or deterioration on the drone's chassis, appendages, and landing gear.

- **Propellers:** Examine propellers for chips, cracks, or any other damage. Verify that they are firmly affixed.
- **Batteries:** Examine batteries for indications of swelling, leaking, or damage. Verify that they are completely charged and correctly inserted.
- **Sensors and Cameras:** Clean the sensors and cameras to guarantee accurate data collection and confirm their proper functionality.
- **GPS and Compass Calibration:** Confirm the calibration of the GPS and compass, and ascertain their proper functionality.

Software and Firmware Updates

Ensure that the drone's software and firmware are current to prevent malfunctions, crashes, or navigational problems during operation.

Battery and Power Levels

Verify the battery levels of the drone and the remote controller. Consistently verify that both are adequately charged for the anticipated flight time.

Flight Zone Safety

Establish Clear Boundaries

- **Designated Flight Area:** Consistently construct a designated flight area devoid of individuals, animals, and obstructions. Establish a boundary that prevents unintended entry into the flight zone.
- **Prohibited Zones:** Refrain from operating in proximity to no-fly zones, including airports, military installations, or regions with significant air traffic. Consistently ascertain local regulations to adhere to flight constraints.

Weather Conditions

1. Refrain from operating drones in adverse meteorological circumstances, including:
2. Intense winds (often exceeding 15-20 mph)
3. Rain, fog, or snow can impair visibility and compromise drone stability.
4. Lightning or thunderstorms, which present considerable hazards to electronic devices.
5. Consistently monitor weather forecasts and verify that conditions are suitable prior to each flight.

Obstacle Awareness

- Remain cognizant of impediments including elevated trees, electrical lines, agricultural edifices, and additional machinery. These may disrupt drone operations and result in mishaps.
- **Unobstructed Flight Paths:** Before flight, evaluate the terrain to ensure the vicinity is devoid of towering obstacles or other impediments that may disrupt the drone's functionality.

Operational Safety During Flight

Line of Sight

Consistently preserve visual contact with the drone throughout its flight. This guarantees that the operator can promptly address any issues or obstacles that may occur.

If operating beyond visual line of sight (BVLOS) is essential, implement supplementary safety measures, including the deployment of spotters or automated fail-safe systems.

Altitude and Speed Limits

Adhere to the prescribed height and speed restrictions established by local aviation authorities. Drones should typically operate at altitudes not exceeding 400 feet above ground level to prevent airspace problems.

Ensure the drone works at a regulated speed and avoids sudden movements that may cause instability.

Emergency Protocols

Establish a comprehensive emergency response plan. This encompasses actions such as:

- **Return to Home (RTH):** Utilize the drone's Return to Home (RTH) feature in the event of signal loss or low battery.
- **Fail-Safe Mechanisms:** Understand the drone's fail-safe protocols, such as landing or maintaining a stationary position in the event of a connection or GPS signal loss.
- **Battery Management:** Continuously monitor battery levels during the flight to guarantee the drone possesses sufficient power for a safe return.

Post-Flight Safety

Safe Landing Practices

Select a level, unobstructed surface for landing to decrease the chance of damaging the drone. Refrain from landing in locations with impediments like as long grass, pebbles, or mud that may hinder the drone's landing gear.

When applying pesticides or fertilizers, ensure the vicinity is devoid of individuals, pets, and livestock during the application procedure.

Equipment Handling

Upon landing, deactivate the drone and controller to avert unintentional activation.

Exercise caution when handling batteries, particularly if they are heated or compromised. Store batteries in a fireproof container or secure location to mitigate overheating or fire hazards.

Post-Flight Inspection

Examine the drone and its components for any indications of wear or damage incurred during the flight.

- If pertinent, cleanse the sensors, propellers, and motors to eliminate dirt, vegetation, or debris.
- **Battery Storage:** Post-flight, ensure the battery is stored appropriately in accordance with the manufacturer's instructions. Refrain from exposing batteries to severe temperatures, and do not store them in fully charged or entirely depleted conditions.

Safety for Operators and Personnel

Operator Training

- Ensure that the drone operator is sufficiently trained in safe flying operations and comprehends the drone's capabilities, constraints, and local restrictions.
- **Certification:** In certain areas, operators may be required to obtain certification to use drones for business purposes. Guarantee adherence to applicable aviation authorities and certification standards.

Personal Protective Equipment (PPE)

When utilizing drones for spraying operations (e.g., pesticides or fertilizers), ensure that operators and adjacent personnel don suitable personal protection equipment, including gloves, goggles, and protective attire.

Maintain a Safe Distance

Consistently keep a safe distance between the drone and individuals, livestock, and wildlife. This mitigates the danger of accidents, including potential injuries from rotating propellers or chemical exposure during application.

Compliance with Legal Regulations

Local Regulations

- Acquaint oneself with local statutes and regulations pertaining to drone operation. This encompasses comprehension of no-fly zones, height limitations, and airspace laws.
- **Flight Authorization:** In certain areas, operators may be required to get authorization for specific flight types (e.g., BVLOS operations or commercial uses).

Privacy Considerations

Guarantee that drone operations uphold individual privacy rights. Refrain from flying over private property without authorization, particularly while utilizing cameras or sensors that gather data from the ground.

Reporting Incidents

Implement protocols for the reporting of accidents, malfunctions, or safety violations. This enhances safety protocols and guarantees adherence to regulatory requirements.

UNDERSTANDING AIRSPACE REGULATIONS AND RESTRICTIONS

Comprehending airspace regulations and constraints is essential for the safe and lawful operation of agricultural drones in India. The nation has instituted explicit regulations to govern drone operations, particularly due to the increasing utilization of drones in agriculture for activities such as crop monitoring, spraying, and soil analysis. These regulations aim to safeguard both drone operators and the public, while also encouraging responsible drone utilization for agricultural and commercial applications.

This document outlines the principal airspace laws and restrictions that agricultural drone operators in India are required to follow.

DGCA (Directorate General of Civil Aviation) Guidelines

The DGCA is the regulatory authority in India tasked with supervising drone operations. It has promulgated norms applicable to diverse drone operations, including agricultural applications.

Key Requirements for Agricultural Drone Operations

- **Drone Categories:** Drones in India are categorized according to their weight.
- **Nano drones:** Weighing 250 grams or less.
- **Micro drones:** weighing between 250 grams and 2 kilos.
- **Small drones:** 2 to 25 kg.
- **Medium drones:** 25 to 150 kg.
- **Large drones:** Exceeding 150 kg.

Agricultural drones generally belong to the small and medium categories, equipped with payloads for crop-spraying apparatus or sensors.

Operational Guidelines

- **Remote Pilot License (RPL):** Drone operators are required to hold an RPL to conduct commercial drone operations, especially in the agricultural sector. The license is granted by DGCA-authorized training companies upon the completion of a requisite training program.
- **Drone Registration:** All drones, excluding nano drones, are required to be registered with the DGCA. This encompasses supplying information regarding the drone, its function, and the operator.
- **Remote Pilot Training:** Operators are required to undergo training to comprehend the concepts of safe drone operation, emergency protocols, and regulatory compliance.

- **Permission for Aerial Work (PFAW):** A PFAW is necessary for drone operations in some instances, such as agricultural applications that include crop spraying or extensive monitoring.

Airspace Restrictions and No-Fly Zones

Agricultural drone activities must comply with the airspace limits and no-fly zones established by the Indian government to ensure safety and adherence to aviation rules.

No-Fly Zones in India

- **Airports and Airstrips:** Drones are forbidden from operating within a 5 km radius of airports or airstrips (public and military) unless explicit authorization is obtained from the DGCA. These zones are essential for ensuring safety in aviation traffic.
- **Military Areas:** Operating drones in proximity to or above military installations or sensitive locations (e.g., defense bases, government edifices) is unequivocally forbidden.
- **Urban and High-Traffic Areas:** Drones should refrain from operating over densely populated urban areas or high-traffic zones to mitigate the potential for mishaps. The DGCA advises against traversing densely populated urban areas.
- **Critical Infrastructure:** Drones must not be operated over critical infrastructure, including power plants, water reservoirs, or other locations that could jeopardize public safety or national security.

Controlled Airspace

- **Class G Airspace:** Drones are often permitted to operate in Class G airspace (uncontrolled airspace) without requiring air traffic control authorization. Operators must maintain vigilance regarding prohibited zones inside this airspace.
- **Class E and C Airspace:** certain zones constitute controlled airspace, necessitating drone operators to obtain prior authorization for flight in certain locations. Operators in agricultural applications must be cognizant of restricted zones surrounding major urban areas or airports.

Aviation Safety Zones

Drones are generally prohibited from flying above 400 feet (120 meters) in most regions unless special authorization is obtained for operations exceeding this altitude. Agricultural drones must remain inside this limit to prevent interference with piloted aircraft and other drones.

Daytime Operations and VLOS (Visual Line of Sight)

The DGCA regulations stipulate that drones are to be flown exclusively during daylight hours and within Visual Line of Sight (VLOS), requiring the operator to maintain visual contact with the drone throughout its flight without the assistance of binoculars or other equipment. This facilitates enhanced control over the drone and permits prompt corrective measures in the event of emergencies.

Exceptions to VLOS

Drones operating outside Visual Line of Sight (VLOS) or in low visibility conditions, such as fog or nighttime, are subject to extra restrictions and necessitate specific authorization from the DGCA.

Drone Flight Permissions and Approvals

Agricultural drone operators are need to secure specific authorizations for particular activities, particularly for intricate operations such as crop spraying or extensive data collection.

Aerial Work Permission (PFAW)

Operators engaged in agricultural activities, such as crop spraying, must have a Permission for Aerial Work (PFAW) from the DGCA. This encompasses the submission of the flight plan, drone specifications, and safety standards.

Flight Plan Submission

Drone operators are required to file a flight plan for operations, with elements such as:

- Flight path, duration, and altitude.
- The objective of the flight (e.g., crop monitoring, spraying, etc.).
- Information regarding any no-fly zones or restricted areas along the flight route.

Drone Insurance and Liability

Although not mandatory, insurance is strongly advised for agricultural drone operators. Insurance policies may encompass potential costs resulting from accidents, including crop damage or personal harm, and may also provide coverage for drone malfunctions or crashes.

Compliance with Data Privacy and Environmental Guidelines

Apart from aviation restrictions, agricultural drone operators in India must comply with environmental and data privacy requirements when gathering data for agricultural activities.

Data Privacy

Drones outfitted with cameras and sensors are capable of capturing images and data pertaining to private assets. Operators must guarantee that drone operations do not violate privacy rights, especially while traversing private properties.

Environmental Impact

Agricultural drones, particularly those utilized for chemical spraying, must adhere to environmentally sustainable standards. Operators must guarantee the responsible application of the chemical payload to prevent runoff or drift into adjacent areas, thereby protecting agriculture, wildlife, and water sources.

Enforcement and Penalties

Noncompliance with airspace regulations and DGCA directives may incur fines, including:

- Penalties for using drones without appropriate registration or authorization.
- Suspension or revocation of operating licenses for recurrent infractions.
- Litigation in instances of negligence or safety violations, including causing accidents or operating in prohibited zones.

Conclusion

Compliance with airspace regulations and restrictions is crucial for the safe and lawful operation of agricultural drones in India. By comprehending and adhering to the directives issued by the DGCA, operators may guarantee the safety of their operations while also facilitating the responsible incorporation of drones in Indian agriculture. Consistently examining updates from the DGCA and remaining cognizant of alterations in the regulatory framework is crucial for ensuring compliance and circumventing potential fines.

EMERGENCY PROCEDURES AND RISK MITIGATION STRATEGIES

Emergency protocols and risk mitigation methods are essential for guaranteeing the safe operation of agricultural drones in India. With the growing utilization of drones for activities such as crop monitoring, spraying, and soil analysis, operators must be equipped to address unforeseen challenges that may occur during flight. The erratic characteristics of agricultural settings, coupled with the technological difficulties of drone operations, necessitate the establishment of comprehensive emergency protocols. Presented below are essential emergency protocols and risk mitigation measures for agricultural drone operators in India.

Pre-Flight Risk Mitigation Strategies

Equipment Inspection and Maintenance

- **Pre-Flight Checklist:** Prior to each flight, operators must do comprehensive inspections of the drone, batteries, sensors, and ancillary equipment. This encompasses verification of:
  - Physical impairment to the drone (e.g., propeller fractures, structural damage).
  - Calibration of GPS and other sensors.
  - Assess battery levels and integrity, ensuring the absence of swelling or leaking.
- **Firmware and Software Updates:** Verify that the drone's software and firmware are current to reduce the likelihood of software failures during operation.

Weather Forecasting

- **Weather Conditions:** Always verify weather conditions before to flying to prevent operating in adverse weather, including strong winds, precipitation, fog, or thunderstorms. Such conditions may result in diminished drone stability, sensor failures, or an elevated risk of accidents.
- **Wind Speed:** Wind gusts exceeding 15-20 mph pose a risk to the majority of agricultural drones. Flight operations must be halted if wind speeds surpass these thresholds.

In-Flight Emergency Procedures

Loss of GPS Signal

- **Symptoms:** A loss of GPS signal may result in instability or deviation from the intended flight route of the drone.
- Emergency Procedure:
  - **Return to Home (RTH):** Numerous drones are equipped with an autonomous Return to Home function that instructs the drone to return to its launch location upon loss of GPS signal.
  - **Manual Control:** In the absence of RTH, the user must endeavor to manually regain control and safely land the drone in an unobstructed region. When the drone operates beyond visual line of sight (BVLOS), employ a drone spotter to assist in locating and directing the drone.
  - **Risk Mitigation:** Confirm that the drone operates within a robust GPS signal range, particularly in locations with elevated structures or dense vegetation.

Battery Failure or Low Battery Warning

- **Symptoms:** Drones often emit a warning when the battery is depleted, potentially resulting in an unanticipated crash if not rectified.
- Emergency Procedure:
  - **Safe Landing:** If Return to Home (RTH) is unfeasible, manually maneuver the drone to land in a secure, unobstructed location, ideally within visual range, to prevent harm to the drone and the surrounding environment.
  - **Risk Mitigation:** Continuously monitor battery levels throughout flight and verify that the drone possesses adequate power for the return trip. Refrain from operating the drone until the battery attains critical levels.

Loss of Communication with the Remote Controller

- **Symptoms:** Should the drone lose connectivity with the remote controller, it may exhibit erratic flight patterns or remain stationary.
- Emergency Procedure:

- **Fail-Safe Mode:** The majority of drones are equipped with inherent fail-safe protocols that activate upon loss of communication, such as hovering or landing.
- **Return to Home:** Should the fail-safe mode fail to rectify the problem, the drone often utilizes GPS coordinates to navigate back to its home point or a designated place.
- **Risk Mitigation:** Ensure that the remote controller remains within the drone's operational range and refrain from flying in proximity to known sources of interference, such as power lines or other wireless equipment.

Post-Flight Emergency Procedures

Drone Crash or Forced Landing

- **Symptoms:** A crash or forced landing may result from mechanical faults, unforeseen weather alterations, or inadequate handling.
- Emergency Procedure:
  - **Evaluate the Damage:** Following a collision or emergency landing, inspect the drone for any apparent damage. Verify that the vicinity is secure for access (e.g., absence of live wires or toxic substances).
  - **Recovery:** If the drone remains intact, promptly power it down and retrieve it from the landing place. Assess the repair requirements of the damaged drone and notify the relevant authorities if warranted (e.g., damage to agricultural crops, property, or infrastructure).
  - **Risk Mitigation:** Consistently prepare for emergency landings by designating open, obstacle-free locations as your landing zones. Refrain from functioning in proximity to essential infrastructure.

Risk Mitigation Strategies for Specific Agricultural Applications

Crop Spraying (Pesticides/Fertilizers)

- **Chemical Exposure:** Improper application of pesticides or fertilizers presents hazards to operators, adjacent workers, and the environment.
- Risk Mitigation:
  - **PPE for Operators:** Operators must use suitable Personal Protective Equipment (PPE), including gloves, goggles, and protective attire, during spraying activities.
  - **Spray Zones:** Avoid operating drones over inhabited regions or in proximity to critical water sources to mitigate chemical drift.
  - **Weather Conditions:** Weather Conditions: Avoid spraying during high winds or precipitation to minimize chemical drift to unexpected locations.

Nighttime or Low Visibility Flights

- **Risk:** Operating an aircraft during nocturnal hours or in situations of reduced visibility heightens the likelihood of crashes, loss of control, or failure to detect obstructions.

- Risk Mitigation:
    - **Avoid Night Flights:** Refrain from operating drones in darkness or low visibility unless they are specifically designed for nighttime use.
    - **Night Flight Authorization:** If nocturnal activities are required, obtain appropriate approval from the DGCA and equip the drone with illumination and night vision capabilities if accessible.
    - **Employment of Spotters:** During nocturnal flights, the presence of a ground spotter can markedly diminish dangers by assisting the operator in sustaining visual contact with the drone.

Training and Preparedness

Operator Training

- **Drone Pilot Certification:** Verify that drone operators possess certification from the DGCA and are proficient in emergency protocols, equipment management, and the utilization of emergency capabilities such as Return to Home (RTH).
- **Scenario Simulations:** Perform routine simulations of emergency scenarios, such GPS signal loss, battery failure, or communication disruption, to enable operators to respond with composure and efficiency under duress.

Emergency Action Plans

- **Flight Plans:** Develop comprehensive flight plans encompassing potential emergency situations and corresponding responses. This enables operators to remain vigilant and diminishes decision-making duration in a crisis.
- **Emergency Contacts:** Compile a directory of emergency contacts for technical assistance, local authorities, or agricultural insurance providers in the event of an accident or equipment malfunction.

Regulatory Compliance and Reporting

DGCA Reporting

- **Accident Reporting:** Operators must report any accident or drone malfunction to the DGCA in accordance with the prescribed procedures. This entails supplying information regarding the occurrence, including the cause, extent of damage, and any injuries sustained.
- **Adherence to Local Legislation:** Guarantee that drone activities conform to all local statutes, including no-fly zones, altitude limitations, and agricultural protection rules.

Emergency protocols and risk reduction methods are essential for the safe operation of agricultural drones in India. Operators can mitigate the probability of accidents and manage emergencies efficiently by anticipating potential risks, performing routine equipment inspections, and providing adequate training. Adhering to regulatory standards and upholding safety protocols safeguards the drone, operators, and environment while enhancing the efficiency and productivity of agricultural drone applications.

12

Successful Drone Applications in Agriculture

Drones have transformed the agriculture sector, altering farming methods by offering advanced solutions for crop monitoring and management. These unmanned aerial vehicles (UAVs) have several applications that improve production, efficiency, and sustainability. Drones help precision agriculture by allowing farmers to collect real-time data on crop health, soil conditions, and field variability. Drones utilize high-resolution imagery and remote sensing technology to identify early signs of pests, illnesses, and nutrient deficits, facilitating targeted interventions. Drones also facilitate irrigation control, seed sowing, and crop spraying, thereby decreasing labor expenses and mitigating environmental impact. Consequently, drone technology is enabling farmers to make better informed decisions, improve resource utilization, and eventually enhance yields.

REAL-WORLD EXAMPLES OF SUCCESSFUL DRONE APPLICATIONS IN AGRICULTURE

Drones are increasingly being used in agriculture across India to boost productivity, ensure efficient resource use, and reduce environmental impact. In recent years, the Indian agricultural sector has witnessed significant adoption of drone technology, which has proven to be a valuable tool for precision farming. Below are some real-world examples of successful drone applications in Indian agriculture:

Precision Farming for Crop Monitoring

In the states of Maharashtra and Punjab, drones are being used by farmers for precision farming to monitor crop health and growth. The drones, equipped with multispectral sensors, capture detailed images of fields and analyze data on crop health, soil moisture, and pest infestations. One example is AeroFarms, an Indian startup that uses drones to monitor fields and help farmers detect early signs of crop diseases and nutrient deficiencies. By assessing the data from drone imaging, farmers can take targeted actions like adjusting irrigation, applying fertilizers, or spraying pesticides, leading to better yields and reduced chemical use.

Pesticide and Fertilizer Spraying

In Andhra Pradesh, drones are revolutionizing the way farmers apply fertilizers and pesticides. The use of drones, like those deployed by Agnix Drone Solutions, allows for precise spraying, reducing wastage and ensuring uniform coverage. This not only saves time but also minimizes human exposure to harmful chemicals. With the help of drones, farmers can spray fertilizers and pesticides on large fields in a matter of hours, compared to traditional methods which could take several days. This is especially beneficial in regions with large farms and challenging terrains, where manual spraying is labor-intensive and inefficient.

Irrigation Management

Water management is a critical issue in Indian agriculture, especially in water-scarce regions. Drones, when equipped with thermal imaging and multispectral sensors, are being used to monitor water distribution in irrigation systems. A notable example is the work done by AeroMegh, which has helped farmers in Rajasthan by using drones to survey water levels in crops. The drone data allows farmers to detect dry patches and areas where irrigation is insufficient, enabling them to optimize water use. By targeting irrigation to areas that need it most, farmers conserve water, improve crop health, and increase overall yields.

Field Mapping and Soil Health Assessment

In states like Tamil Nadu and Uttar Pradesh, drones are being used to map fields and assess soil health. Pixxel, an Indian satellite imaging company, provides agricultural drone services that map soil conditions across vast areas. By identifying areas with poor soil quality or low nutrient levels, farmers can adjust their agricultural practices, such as changing crop rotation patterns or adding specific fertilizers. This level of soil insight ensures that farmers can make data-driven decisions to maximize crop productivity while maintaining sustainable farming practices.

Livestock Monitoring

In dairy farming, drones are being used to monitor livestock health and behavior. In places like Haryana, farmers have started using drones to track the movement and condition of their cattle, especially in large dairy farms. Drones can quickly cover large areas of pasture, identifying signs of disease or distress in livestock before they become serious problems. This has improved the efficiency of farm management by reducing the need for extensive manual inspections.

Disaster Management and Crop Loss Assessment

Drones have also been effective in post-disaster assessments. In the aftermath of floods, cyclones, or droughts, drones are used to assess crop damage. For instance, after Cyclone Fani hit Odisha in 2019, drones were deployed to assess crop losses and help farmers understand the extent of the damage. This helped the government and insurance companies in faster claims processing and also in deciding relief strategies for affected farmers.

In India, the application of drones in agriculture is an emerging and transformative trend that is making farming more efficient, sustainable, and cost-effective. Through precision farming, pesticide spraying, irrigation management, field mapping, and livestock monitoring, drones have provided Indian farmers with powerful tools to improve productivity, reduce costs, and conserve resources. As drone technology continues to evolve and become more affordable, its role in Indian agriculture will only grow, leading to more widespread adoption and further advancements in agricultural practices.

Examples of Successful Drone Applications in Agriculture in Indian

In India, drone technology is progressively being utilized in agriculture to improve productivity, efficiency, and sustainability. Numerous effective real-world instances demonstrate the use of drones in revolutionizing agricultural techniques nationwide:

- **Drone-Assisted Pesticide Application in Punjab:** Farmers in Punjab have been using drones for pesticide application, markedly enhancing efficiency and diminishing labor expenses. Drones outfitted with sprayers may efficiently cover extensive areas in a brief period, guaranteeing uniform pesticide distribution and minimizing the potential for crop damage. This approach also reduces farmers' exposure to hazardous chemicals, hence enhancing safety and environmental sustainability.
- **Drones for Precision Agriculture in Karnataka:** In Karnataka, drones are utilized for precision agriculture, namely for monitoring crop health. Agriculturists employ drones to get multispectral imagery, facilitating the identification of regions impacted by diseases, pests, or nutritional deficits. This data facilitates focused actions, diminishes input costs, and enhances overall crop health. This method is particularly advantageous for crops such as sugarcane, cotton, and maize.
- **Maharashtra's Utilization of Drones for Agricultural Surveillance:** In Maharashtra, drones are employed for the surveillance of extensive agricultural lands, especially in the Vidarbha region. Drones assist in monitoring the health of cotton fields by detecting stress indicators and early pest infestations. Agriculturists can utilize this information to administer fertilizers and pesticides with more precision, thereby enhancing productivity and reducing environmental impact. The state government has initiated pilot initiatives to promote the utilization of drones in agriculture.
- **Drones for Horticultural Surveillance in Uttarakhand:** In Uttarakhand, drones are utilized for the surveillance of horticultural crops such as apples and pears. Drones gather comprehensive data on crop health, soil moisture, and insect infestations by surveying orchards from the air. This enables farmers to enhance irrigation, fertilizer application, and insect management, resulting

in healthier crops and increased yields, while simultaneously minimizing resource consumption.

- **Drone-Assisted Water Management in Andhra Pradesh:** Drones are utilized for agricultural water management in Andhra Pradesh. Drones outfitted with infrared cameras facilitate the monitoring of soil moisture levels over extensive agricultural areas, allowing farmers to make more educated irrigation decisions. This method is especially advantageous in arid regions, where effective water utilization is essential for agricultural sustainability.

CASE STUDIES AND SUCCESS STORIES

Notwithstanding the limitations, there are numerous exemplary instances of effective drone utilization in agriculture within rural India:

Crop Monitoring in Maharashtra

Farmers in Maharashtra are use drones with multispectral cameras to assess crop health, identify nutrient deficiencies, and enhance fertilizer application, resulting in increased yields and enhanced resource efficiency.

Pest Management in Punjab

Drones are employed in Punjab to detect pests, particularly the destructive locusts, and to administer precise pesticide applications, therefore decreasing reliance on broad-spectrum poisons and mitigating environmental consequences.

HANDS-ON EXERCISES AND FIELD DEMONSTRATIONS

Over the past two years, farmland has undergone a transformation through innovative crop monitoring techniques, precision agriculture drones, and resource optimization facilitated by agricultural drones. A hands-on training and demonstration session was arranged to inform the farming community and its fans about the proper use of modern instruments.

Practical exercises and field demonstrations are crucial elements for the effective integration of drones into agricultural activities. They enable farmers and agronomists to engage with drone technology in practical situations and comprehend its complete capacity for improving productivity. Below are few instances of exercises and demonstrations that have been well executed in agriculture:

Drone Training Workshops in Punjab

- **Objective:** To instruct farmers and agronomists on the operation of drones for applications including crop spraying, health assessment, and field mapping.
- **Activities:** The session features live demonstrations of drones applying pesticides and fertilizers to crops. Participants receive instruction on programming drones, interpreting aerial imagery, and analyzing data for agricultural management. Practical exercises enable participants to operate drones under supervision, providing them with experiential knowledge.

- **Impact:** Farmers in the region have adopted drones to enhance pesticide application, minimize labor expenses, and improve operational efficiency.

Karnataka Precision Agriculture Pilot Project

- **Objective:** To illustrate the role of drones in precision agriculture through the acquisition of multispectral imagery of crops for analytical purposes.
- **Activities:** Field demonstrations entail operating drones over agricultural fields to gather photographs and data regarding crop health, soil moisture, and insect infestation. Participants can observe how drones identify agricultural stress, enabling them to strategize solutions more efficiently. The data gathered throughout the flight is evaluated with agricultural software to provide relevant insights.
- **Impact:** Farmers get knowledge on utilizing drones for precise interventions such as fertilization, irrigation, and pest management, leading to cost reductions and increased agricultural yields.

Maharashtra Drone Demonstrations for Cotton Farms

- **Objective:** To educate farmers in Maharashtra on the utilization of drones for monitoring cotton crops and identifying early indicators of illnesses or pest infestations.
- **Activities:** Demonstrations occur on cotton farms, utilizing drones to survey extensive fields and illustrate how high-resolution imagery can identify problematic regions. The experiment additionally involves employing drones to apply pesticides to designated locations informed by the health data collected by the drones.
- **Impact:** Farmers acquire insights into the efficacy of drone-assisted monitoring and pesticide application, resulting in time savings and less chemical usage.

Uttarakhand Horticulture Drone Demonstrations

- **Objective:** To demonstrate the utility of drones in the surveillance of apple and pear orchards situated in mountainous areas.
- **Activities:** Participants in field demos witness drones operating over orchards to acquire high-resolution images that assists in identifying crop health, pest infestations, and watering requirements. They acquire practical experience utilizing software to evaluate drone data and generate field maps for enhanced decision-making.
- **Impact:** Horticulturists recognize that drones can enhance yield management and diminish the necessity for manual checks, thereby expediting and refining the process.

Andhra Pradesh Water Management with Drones

- **Objective:** To illustrate the use of drones in enhancing water efficiency for agricultural irrigation.

- **Activities:** Drones outfitted with infrared sensors traverse fields to collect soil moisture data. Participants are instructed on how to analyze these photos to identify regions of the field requiring more or decreased irrigation. Practical exercises enable learners to configure and operate drones for the surveillance of crops across various irrigation zones.
- **Impact:** Farmers in arid locations acquire expertise in optimizing water supplies with drone technology, thereby promoting sustainable agricultural methods.

Key Benefits of These Demonstrations

- **Practical Knowledge:** Participants get experiential understanding of drone operation, enhancing their comprehension of the technology's possibilities and limits.
- **Skill Development:** Farmers acquire proficiency in the proper utilization of drones for targeted agricultural functions such as spraying, monitoring, and data analysis, hence enhancing their efficiency and productivity.
- **Enhanced Adoption:** Observing the tangible advantages of drones motivates farmers to embrace these technology, resulting in superior agricultural practices.
- **Customized Solutions:** These demos can be adapted for various crops, climates, and agricultural conditions, guaranteeing that the solutions are pertinent and applicable to the local context.

Integrating practical exercises and field demonstrations into agricultural training programs enables farmers to effectively utilize drone technology, hence enhancing farm productivity and sustainability.

Example

The seminar commenced with an instructive presentation regarding several types of agricultural drones, their functionalities, and the advantages they offer. Drones can be utilized for crop health assessment, soil monitoring, irrigation management, and further applications, which specialists elaborated on extensively.

Live Demonstration

Following the introduction, the audience observed a live demonstration of drone operation. A certified pilot exhibited the procedure for configuring a drone for flight, including pre-flight inspections, takeoff, and data acquisition. The participants observed a drone obtaining high-resolution photos and near-real-time data regarding crop conditions.

Hands-on Training

Following the show, the participants operated the drones under the guidance of qualified trainers, receiving instruction on flight navigation, data collection, and safety protocols. They all engaged in practical experience by implementing them in real operations without prior knowledge of the actual tasks involved.

Data Analysis Workshop

The capacity to analyze data gathered by drones is essential for effective decision-making in agriculture. A workshop was held on the analysis of data collected by drones utilizing software tools. The trainees acquired skills in map creation, crop health assessment, and the formulation of practical insights derived from aerial images.

Q&A Session

The event concluded with a question-and-answer session, during which participants engaged with the experts. Inquiries encompassed operational difficulties and regulatory factors, guaranteeing that participants departed the workshop with a thorough comprehension of how to integrate drone technology into their respective professions.

The practical training and display of agricultural drones effectively enhanced comprehension of technology's potential to transform farming. Participants, equipped with enhanced skills and knowledge, were ready to use drones into their agricultural activities. Such events are crucial for the industry's advancement, as they empower farmers to spearhead the integration of novel technologies for enhanced productivity and long-term sustainability.

13

Widespread Adoption of Drone Technology in Agriculture

The extensive integration of drone technology in agriculture is swiftly revolutionizing farming practices, providing substantial enhancements in efficiency, precision, and sustainability. Drones provide farmers a cost-efficient and effective method for crop monitoring, resource management, and yield optimization. Drones facilitate high-resolution imagery acquisition, real-time crop health monitoring, and precise application of fertilizers or pesticides, empowering farmers to make data-driven decisions that improve productivity. As technology becomes increasingly accessible, particularly in developing regions, its usage is steadily rising. Agriculturalists are progressively recognizing the capacity of drones to decrease labor expenses, mitigate resource wastage, and improve environmental sustainability, establishing them as a crucial instrument in the future of agriculture.

CHALLENGES AND OPPORTUNITIES FOR THE WIDESPREAD ADOPTION OF DRONE TECHNOLOGY IN AGRICULTURE

The extensive integration of drone technology in agriculture offers both obstacles and prospects. Comprehending these factors can influence the future of agriculture and realize its complete potential. Drone technology is being integrated into Indian agriculture; yet, obstacles to wider implementation persist.

Challenges

- **Significant Initial Expense:** The initial capital required for acquiring drones, along with essential sensors and software, can be excessive, particularly for small-scale agricultural producers. The expenses associated with training and upkeep exacerbate the financial strain, hindering certain farmers from embracing the technology.
- **Technical Skills and Training:** Proficient operation of drones necessitates a specific degree of technical acumen. Agriculturalists must acquire the skills

to operate drones, analyze the obtained data, and interpret it with precision. The deficiency of adequate training and proficient staff in rural regions can impede extensive adoption.

- **Regulatory Challenges:** In numerous countries, drone operations face stringent laws, including prohibitions on flying drones above specific altitudes or in proximity to populated zones. Agricultural producers must maneuver through intricate legal structures, and in certain regions, regulatory authorities have not completely embraced drone utilization in farming, resulting in ambiguity.
- **Data Overload:** Drones can provide extensive volumes of data, which may be daunting for farmers lacking the necessary tools and skills to analyze and interpret it. Transforming this data into useful insights necessitates specialized tools and expertise, which may pose a barrier for many.
- **Weather Conditions:** Drones are significantly reliant on optimal weather conditions for efficient functioning. Wind, precipitation, and severe temperatures might impede their effectiveness or perhaps destroy the equipment, restricting their utilization during specific seasons.
- **Cost and Affordability:** The initial expenses associated with acquiring drones and other equipment, together with the training of staff for operation and maintenance, might be excessively burdensome for smallholder farmers and resource-limited communities in rural India, hence restricting their access to this technology.
- **Regulatory Obstacles:** The utilization of drones in agriculture is governed by regulations and restrictions established by civil aviation authorities, encompassing airspace regulations, licensing prerequisites, and limitations on flight operations, which may present challenges for farmers aiming to implement drones for agricultural applications.
- **Insufficient connection and Infrastructure:** Numerous rural regions in India are deficient in essential connection and infrastructure, such as electricity, internet access, and drone landing places, obstructing the deployment and operation of drones in these locales.
- **Data Management and Privacy:** The collection, processing, and analysis of data obtained from drones necessitate specialized expertise, tools, and infrastructure, alongside protocols to safeguard data privacy, security, and confidentiality, hence prompting issues over data ownership, misuse, and ethical implications.

Opportunities

- **Increased Efficiency and Precision:** Drones enhance efficiency and accuracy, allowing farmers to assess crop health, soil conditions, and field variability precisely, resulting in more focused treatments. This precision agriculture minimizes the excessive application of fertilizers, pesticides, and water, resulting in cost reductions and improved crop yields.

- **Access to Real-Time Data:** Drones provide farmers with real-time insights into their fields, enabling prompt and informed decision-making. This is particularly advantageous for overseeing extensive regions, where conventional techniques may be excessively time-consuming and laborious.
- **Enhanced Resource Management:** Drones facilitate effective irrigation, optimum planting, and accurate pesticide application, thereby conserving resources such as water and chemicals. By utilizing these resources more efficiently, farmers may enhance sustainability and mitigate environmental damage..
- **Scalability and Flexibility:** Drones offer scalability and flexibility, applicable to many farm sizes, ranging from small holdings to big industrial farms. The method is adaptable to several crop varieties, rendering it highly versatile. With the ongoing decline in drone pricing, even small-scale farmers can begin to reap the advantages of this technology.
- **Data-Driven Decision Making:** The data gathered by drones may be evaluated to anticipate trends, project crop performance, and evaluate field conditions. This enables farmers to obtain meaningful insights for proactive decision-making, including the adjustment of planting schedules, management of pest outbreaks, and optimization of irrigation systems.
- **Government Assistance and Innovation:** In numerous nations, including India, governments are providing subsidies, grants, and policy support to encourage the adoption of drone technology in agriculture. Furthermore, continuous advancements in drone design, artificial intelligence, and machine learning are rendering the technology increasingly accessible, economical, and user-friendly.

In summary, although the extensive implementation of drone technology in agriculture encounters numerous obstacles, these can be alleviated by investments in education, infrastructure, and legal frameworks. Conversely, the advantages presented by drones—such as precision agriculture, resource conservation, and data-informed decision-making—possess the capacity to transform agriculture, enhance production, and foster a more sustainable future.

ETHICAL AND SOCIETAL IMPLICATIONS OF DRONE USE IN FARMING

The utilization of drones in agriculture raises various ethical and societal concerns, such as privacy violations, possible environmental repercussions from pesticide application, data security, equitable technology access, and the necessity for explicit regulations to guarantee responsible usage, especially concerning the collection and utilization of data obtained by drones on private land.

The utilization of drones in agriculture presents numerous ethical and societal ramifications that require thorough examination. The technology presents considerable benefits for efficiency, sustainability, and production; yet, its extensive implementation also prompts worries about privacy, employment, environmental consequences, and equity. Resolving these difficulties is essential to guaranteeing that drone technology serves society comprehensively, without intensifying current challenges or generating new ones.

Ethical Implications

- **Privacy Issues:** Drones outfitted with high-resolution cameras and sensors can record intricate photos and data of agricultural areas, occasionally surpassing mere crop surveillance. The capacity to inadvertently observe adjacent territories or residential zones engenders apprehensions regarding privacy. Farmers must be cognizant of the limits of their observation, ensuring they do not violate the privacy of others while observing their crops.
- **Data Ownership and Security:** Drones amass extensive data concerning crop health, soil conditions, and environmental variables. Questions emerge over the ownership of this data—specifically, whether it belongs to the farmer, the drone service provider, or the technology manufacturer. Furthermore, there are apprehensions regarding data security, as the information gathered may be susceptible to hacking or illegal access. It is imperative to guarantee secure storage and transparent data ownership policies.
- **Environmental Impact:** Although drones are frequently regarded as more eco-friendly than conventional farming techniques, the materials utilized in their production, like batteries and plastics, may pose environmental risks. If drones are not adequately disposed of or repurposed, they may exacerbate pollution. The augmented utilization of drone technology may result in elevated energy usage and an enhanced carbon footprint if not controlled judiciously.

Societal Implications

- **Job Displacement:** A primary concern associated with drone technology is its capacity to displace laborers in the agricultural industry. Manual tasks, including crop monitoring, pesticide application, and field mapping, can now be executed by drones, potentially diminishing the demand for labor. Although drones may generate new employment in technology and data analysis, there exists a possibility that numerous low-skilled agricultural workers may face unemployment if they are not re-skilled.
- **Access and Equity:** The implementation of drone technology may exacerbate the digital divide, since larger, affluent farmers and agribusinesses may afford and leverage the technology, while smallholder farmers or those in poor nations may be marginalized. To guarantee equitable access, it is essential to offer affordable drone solutions and training to marginalized people, especially in rural regions where conventional farming practices prevail.
- **Rural Communities and Technological Reliance:** Although drones can enhance production, there exists a risk of fostering dependency on technology, particularly if farmers become reliant on drones for all facets of farm management. This may provide challenges in areas with inadequate technological infrastructure or where drone maintenance and repairs are not readily available. Rural communities may experience interruptions due

to technological failures, rendering them susceptible to market volatility or natural calamities.

- **Cultural Shifts and Traditional Farming Practices:** The introduction of drones may result in a departure from conventional farming practices, typically transmitted between generations. In certain instances, drone technology may conflict with cultural or local agricultural practices, resulting in possible societal discord. It is essential that technology combines with and enhances traditional practices instead of completely supplanting them, in order to uphold cultural identity and conserve agricultural legacy.
- **Environmental Monitoring and Conservation:** Drones have considerable promise for environmental monitoring and conservation. They can be utilized to evaluate land deterioration, monitor water consumption, and assess biodiversity. Drones could significantly contribute to environmental stewardship and ecosystem protection by facilitating the adoption of more sustainable practices among farmers, particularly in areas threatened by deforestation or soil erosion.

Balancing Ethics and Opportunities

To alleviate the adverse ethical and societal consequences of drone utilization in agriculture, stakeholders including governments, agricultural organizations, and technology firms must collaborate to:

- Enact regulations that mitigate privacy issues and guarantee the transparent and responsible utilization of drones in agriculture.
- Promote training and skill enhancement for workers potentially affected by automation, guaranteeing that new employment possibilities in technology and drone maintenance are available to rural communities.
- Promote equitable access to technology, ensuring that small-scale farmers possess the tools and assistance necessary to incorporate drones into their operations.
- Promote sustainability through the development of environmentally friendly drones and the advocacy of recycling and waste management to mitigate the ecological impact of drone manufacturing and utilization.

Drones in agriculture possess significant potential to enhance farming operations; nonetheless, their ethical and sociological ramifications must be meticulously addressed to guarantee responsible and equitable usage. By mitigating these issues, drone technology can enhance a more sustainable, efficient, and equitable agricultural future.

Bibliography

Agarwal, N., & Jain, P. (2019). Drone Applications for Soil Monitoring in Indian Agriculture. Journal of Agricultural Research and Innovations, 29(3), 91-98.

Agri Innovations. (2021). Drone Technology and Its Role in Indian Agriculture. Retrieved from http://www.agriinnovations.com/india-drones

Agribotix. (2021). Drone-based Solutions for Indian Agriculture: Applications and Maintenance. Retrieved from http://www.agribotix.com/india

Agricultural Exporters Association. (2021). Drones for Monitoring and Crop Inspection in India: Technological Challenges. AEAI Report.

Agri-Drones Monthly. (2020). Essential Maintenance Practices for Agricultural Drones in Rural India. Agri-Drones Monthly.

Agri-Drones Tech. (2020). How to Operate Agricultural Drones Efficiently: A Guide for Indian Farmers. Agri-Drones Tech Journal.

Agri-Tech India. (2021). Drone Technology Adoption in Rural India: Challenges in Maintenance and Usage. Agri-Tech Report.

Agri-Tech India. (2021). The Role of Drones in Indian Agriculture and Their Maintenance. Agri-Tech Report.

Agro Technology. (2020). Best Practices for Drones in Crop Health Monitoring. Agro Technology Journal.

Arora, A. (2020). Agricultural Drone Systems: Operation, Maintenance, and Management in Indian Farming. Innovators Press.

Arora, D., & Gupta, R. (2021). Drone-Based Pest Control Systems in Indian Agriculture: Operational Efficiency and Maintenance. Agricultural Engineering Journal, 18(1), 88-95.

ASSOCHAM. (2021). Drone Technology in Indian Agriculture: Operational Guidelines and Best Practices. ASSOCHAM Report.

CII. (2021). Drones in Indian Agriculture: Potential, Challenges, and Maintenance Practices. Confederation of Indian Industry Report.

CropLife India. (2020). Drones for Agricultural Pesticide Applications in India: Safety and Maintenance. CropLife India Report.

CropTech India. (2020). Drone Application Technologies in Indian Agriculture: Operational and Maintenance Practices. CropTech India Journal.

DGCA. (2021). Standard Operating Procedures for Drone Maintenance in Indian Agriculture. Government of India.

Directorate General of Civil Aviation (DGCA), Government of India. (2021). Guidelines for the Operation of Drones in Indian Agriculture. DGCA Publications.

Directorate General of Civil Aviation (DGCA), Government of India. (2021). Safety Guidelines for Drone Operations in Indian Agriculture. DGCA Document.

Directorate General of Civil Aviation (DGCA). (2021). Drone Registration and Operational Guidelines in India. Retrieved from http://www.dgca.gov.in/drone-guidelines

Directorate of Plant Protection, Quarantine, and Storage. (2021). Guidelines for Drone Applications in Crop Protection. Government of India.

Drone Ecosystem India. (2021). Guidelines for Drone Operations and Maintenance for Indian Farmers. Retrieved from http://www.droneecosystemindia.com

Drone India Magazine. (2021). Maintaining Agricultural Drones in Indian Environments: A Guide. Drone India Magazine.

Drone India. (2020). Operational and Maintenance Insights for Agricultural Drones in India. Drone India Magazine.

FICCI. (2020). Drones in Precision Agriculture: Future Trends and Maintenance Issues in India. FICCI Report.

FICCI. (2020). The Growth of Drone Applications in Indian Agriculture. FICCI Industry Report.

Food and Agriculture Organization (FAO). (2021). The Role of Drones in Precision Agriculture in India. FAO Report.

Government of India Ministry of Agriculture and Farmers Welfare. (2020). Using Drones for Sustainable Farming: Operation and Maintenance in India. Retrieved from http://www.agriculture.gov.in/drone-agriculture

Government of India. (2020). Regulatory Framework for Drone Usage in Indian Agriculture. Ministry of Civil Aviation.

Government of India. (2021). Standard Operating Procedures for Drone Usage in Agriculture. Ministry of Agriculture.

Gupta, M. (2020). Maintenance Practices for Drones in Precision Agriculture in India. Punjab Agricultural University, M.Sc. Thesis.

Gupta, P. (2019). Technological Advancements in Agricultural Drones. Agro Solutions Press.

Gupta, R., & Joshi, A. (2020). Challenges in Drone Maintenance for Indian Farmers. National Conference on AgriTech Innovations, Pune, 89-95.

Gupta, T., & Patel, S. (2020). Effective Drone Operations and Maintenance for Indian Agriculture. TechnoAgri Books.

Indian Chamber of Commerce. (2020). Challenges and Solutions for Drone Operations in Indian Agriculture. ICC Report.

Indian Council of Agricultural Research (ICAR). (2019). Drones in Indian Agriculture: Research and Development. ICAR Press.

Indian Council of Agricultural Research. (2020). Drone Applications and Support for Indian Farmers. Retrieved from http://www.icar.org.in/drone-operations

Indian Farming Tech. (2021). Drones in Agriculture: Maintenance and Performance in Indian Conditions. Indian Farming Tech Journal.

Indian Institute of Remote Sensing, ISRO. (2021). Drone Applications in Indian Agriculture: A Satellite and Drone Integration Approach. Retrieved from http://www.isro.org/drone-applications

Indian Institute of Technology (IIT), Kharagpur. (2020). Drone Usage for Crop Monitoring and Management in India. IIT Research Report.

Indian Ministry of Agriculture and Farmers Welfare. (2020). Use of Drones in Sustainable Agricultural Practices in India: Policy Guidelines. Government of India.

Indian Ministry of Agriculture and Farmers Welfare. (2021). Policy Framework for Drone Usage in Indian Agriculture. Government of India.

Indian Ministry of Agriculture. (2020). Government Initiatives for Drone Adoption in Indian Agriculture. Retrieved from http://www.agriculture.gov.in/drones

Indian Ministry of Rural Development. (2021). Policy on Agricultural Drones and Maintenance Support. Government of India.

Indian Ministry of Rural Development. (2021). Policy on Drone Technology Integration in Indian Agriculture. Government of India.

Indian Ministry of Rural Development. (2021). Technology-driven Solutions in Indian Agriculture: Drones and Their Applications. Ministry of Rural Development Report.

Indian Ministry of Science and Technology. (2021). Drone Technology for Agricultural Development in India: A National Policy Framework. Government of India.

Indian Ministry of Science and Technology. (2021). National Framework for Drone Research and Applications in Agriculture. Government of India.

Indian Space Research Organisation (ISRO). (2020). Integration of Remote Sensing with Agricultural Drones in India: Policy and Operational Guidelines. ISRO Report.

Jain, A. (2020). Drone Technology in Indian Agriculture: Key Challenges in Operation and Maintenance. National Institute of Technology, Warangal, Ph.D. Dissertation.

Jain, S., & Gupta, R. (2020). Maintenance Strategies for Agricultural Drones in India. Indian Journal of Drone Engineering, 25(1), 67-74.

Joshi, K., & Gupta, M. (2021). Comparative Study of Drone Maintenance Strategies in India. Conference on Agricultural Robotics and Automation, Delhi, 88-94.

Joshi, K., & Mehta, P. (2020). Drones for Agricultural Monitoring: Challenges and Future Directions in India. International Conference on Agricultural Robotics and Drones, Bengaluru, 134-140.

Joshi, M. (2019). Operation and Management of UAVs in Agriculture in India. AgriWorld Publications.

Joshi, S., & Patel, N. (2020). Handbook of Drone Technology and Agricultural Applications in India. AgriTech Publishing.

Kisan Tech India. (2021). How Drones Are Changing the Indian Agricultural Landscape. Retrieved from http://www.kisantechindia.com

Kumar, A. (2020). Drone-Based Agricultural Monitoring and Maintenance Practices in India: Challenges and Opportunities. Indian Agricultural Research Institute, Ph.D. Dissertation.

Kumar, R., & Singh, A. (2020). Drones for Precision Agriculture in India: Current Trends and Future Prospects. Indian Journal of Agricultural Technology, 32(4), 45-52.

Kumar, V. (2020). Drone Applications in Precision Farming in India: Operation and Maintenance Guidelines. AgriVision Publishers.

Kumar, V., & Singh, R. (2021). Trends in Drone-Based Crop Monitoring and Maintenance in India. Journal of Agricultural Robotics, 27(2), 134-141.

Mehta, A. (2020). Agricultural Drone Maintenance and Calibration Systems in India: A Technical Study. University of Rajasthan, M.Tech Thesis.

Mehta, K. (2021). Drone Applications for Sustainable Agriculture in India: A Complete Guide. Kisan Tech Publications.

Mehta, M., & Sharma, R. (2021). Flying High: The Role of Drones in Indian Agriculture. Indian Agricultural Press.

Mehta, N., & Jain, M. (2021). Drone Technologies for Soil Health Monitoring in India: Operational Practices. Journal of Sustainable Farming Technologies, 19(2), 48-54.

Mehta, S., & Verma, K. (2020). Drone-based Spraying Systems for Crop Protection in India. Journal of Pest Control and Agriculture, 14(6), 118-125.

Mehta, V., & Kumar, G. (2020). Operational Efficiency of Drones for Agricultural Surveying in Rural India. International Conference on Drone Engineering and Maintenance, Hyderabad, 64-70.

Phogat, V., & Mahajan, G. (2021). Precision Agriculture: Technologies and Global Markets. Nova Science Publishers.

Rai, M., & Radhakrishnan, R. (2022). Drone Technology in Agriculture: Opportunities and Challenges. ICAR – National Agricultural Higher Education Project (NAHEP).

Singh, A., & Chauhan, R. S. (2021). Agricultural Engineering: Fundamentals and Applications. Biotech Books.

Thenkabail, P. S., Lyon, J. G., & Huete, A. (2018). Hyperspectral Remote Sensing of Vegetation. CRC Press.

Vick, B. D., & Green, O. O. (2016). Introduction to Unmanned Aircraft Systems. CRC Press.

Vosselman, G., & Maas, H.-G. (2010). Airborne and Terrestrial Laser Scanning. CRC Press.

Zhang, C., & Kovacs, J. M. (2012). The application of small unmanned aerial systems for precision agriculture: a review. Precision Agriculture, 13(6), 693–712.

Index